# BIRMINGHAM INVENTORS and INVENTIONS

# BIRMINGHAM
# INVENTORS and INVENTIONS

BEING A CONTRIBUTION TO THE

## INDUSTRIAL HISTORY OF BIRMINGHAM

BY RICHARD B. PROSSER,
H.M. Patent Office

WITH A NEW FOREWORD BY

ASA BRIGGS
Professor of History, Vice-Chancellor of the University of Sussex

REPUBLISHED BY S.R. PUBLISHERS, 1970
ORIGINALLY PUBLISHED FOR PRIVATE CIRCULATION
(fifty copies), BIRMINGHAM, 1881

East Ardsley, Wakefield, Yorkshire, England

ISBN 0 85409 578 0



Reproduced and Printed by Redwood Press Limited,

Trowbridge and London

## FOREWORD

The origins of the patent system have been traced back to early Greek civilisation, but it is only since the advent of developing industrial technology that the economic and social implications of the issuing of patent rights have been considered. Even now, scholarly writing on the subject is patchy, and the analysis of the influence of different kinds of patent systems on industrial progress is incomplete. Richard B. Prosser was not particularly interested in the general bearings of his subject when he produced this valuable little book in 1881 on the basis of articles he had printed in the previous year. It was a pioneering enterprise, but it was informed more by local pride, as the first chapter makes abundantly clear, than by any attempt at economic anlaysis. Prosser worked in the Patent Office, and he had easy access to all available information. Yet he relied mainly, as he says, on his own reading and on personal recollections. There was obviously ample scope in an industrial city like Birmingham for finding things out which did not appear in official papers. His book is brightly written and was greatly appreciated by those people who had access to it. Fifty copies only were printed, and it has been a difficult book to acquire from the very start.

To relate it to national patent history, it is necessary to study the chronological lists of patents applied for and sealed. The national statistics are usefully set out in B. R. Mitchell and Phyllis Deane, 'Abstract of British Historical Statistics' (Cambridge University Press, 1962), pages 268–9. Down to 1852 the figures relate only to England: after a major change in the law in 1852 when the statistics of applications for patents also became available, the figures relate to the United Kingdom. The relevant comparative table in Prosser is to be found on page 227. He has very little to say in the main body of his book, however, about the period after 1830.

For the detailed local history, it is necessary to turn to a number of more recent books on Birmingham's industrial history. Prosser relied for background on S. Timmins, 'The Resources, Products and Industrial History of Birmingham and the Midland Hardware District' (Hardwick, 1886), a series of reports collected by the local industries committee of the British Association when it met in Birmingham in 1865. This volume, packed with information, may be usefully compared with 'Birmingham and its Regional Setting' (1950), a collection of papers and articles prepared for the visit of the British Association to Birmingham eighty-five years later in 1950. In the first volume the emphasis was mainly on the technology and organisation of particular trades: in the second it shifted to 'the effects of the technological changes of the Industrial Revolution upon landscape and social conditions generally'. Prosser would have found the second of these volumes as interesting for him as the first, for the kind of statements of a general kind which he frequently makes in his book (for example, about the role of innovation in Birmingham and Manchester) depend upon an appreciation and an assessment of local factors as a whole and not simply of the rate of technical progress in particular industries.

The other twentieth-century books which constitute necessary reading to set Prosser's material within an adequate frame are W. H. B. Court 'The Rise of the Midland Industries 1600–1838' (Oxford University Press, 1938), G. C. Allen, 'The Industrial Development of Birmingham and the Black Country' (Allen and Unwin 1929) and the Victoria County History of Warwick, Volume 7, 'The City of Birmingham' (Oxford University Press 1964). There is also an important general account of the background of developments of chemistry in A. and N. L. Clow, 'The Chemical Revolution' Batchworth Press (1952). Prosser was less interested in the diffusion of inventions and the lags between initial discovery and general application than modern writers. Nor have all his verdicts been echoed completely by later historians. He is kinder to Lewis Paul, for example, in Chapter II than many who have followed him in tracing Paul's achievements, and because he felt that Matthew Boulton had been dealt with frequently and at length by other writers his account of Boulton is somewhat summary. He had little to say about the fierce fights centering on patent rights and the effect of the fights on the rate of industrial growth. These aspects of the Birmingham story and the cultural 'milieu' in which invention and application were achieved and extended are examined at length in Robert E. Schofield's 'The Lunar Society of Birmingham: A Social History of Provincial Science and Industry in Eighteenth-Century England' (Oxford 1963).

There is one further general point about Birmingham and its very special place in national history which was not clearly made by Prosser. Although it was the home of many of the key inventions in the history of steam power, the businessmen of the city were slow to employ steam power in the wide spectrum of local metal industries, some of them operating in very small workshops. There were reasons for this which lie deep in the economic history of the period. Some of them were set out by G. C. Allen in his important book on industrial development: some of them were noted as early as 1832 by Charles Babbage in his 'Economy of Machinery and Manufactures'. The record of Manchester and Birmingham in relation both to invention and application should not be settled, as Prosser implies, on the basis of local pride: it demands the fullest possible treatment in terms of both economic and social history.

Asa Briggs

The University of Sussex, March 1970.

# Birmingham Inventors

and

# Inventions,

being a contribution to the

## INDUSTRIAL HISTORY OF BIRMINGHAM.


by

RICHARD B. PROSSER,

*H. M. Patent Office.*


[*Fifty Copies only for Private Circulation.*]


BIRMINGHAM:

The "Journal" Printing Works, New Street.

1881.

**Please note :**

Pages iv (continuation of Preface), v and vi appear in the following order :
vi, v, iv

# PREFACE.

The following pages appeared in the *Birmingham Weekly Post* during last year, and they are now reprinted with a few corrections and additions.

The great importance of the series of printed specifications of patents, as the groundwork of the industrial history of a district, or the history of a particular trade, is now generally admitted. The writer has accordingly made large use of that valuable source of information, which has been supplemented by the results of many years' diligent note-taking, and also by his own personal recollections.

As will be seen by a glance at the Table of Contents, the method of treatment has been partly according to subjects, and partly chronological. Every Birmingham patent, from the earliest period down to 1830, has been dealt with. The more important trades have been treated separately, and the history has been brought down to the year 1852, and, in some instances, later. Many unpatented inventions have been noticed, but the difficulty in such cases is to obtain trustworthy particulars of names and dates—a difficulty which does not occur with patented inventions.

An endeavour has been made to clothe the somewhat dry bones of the specifications with flesh, by tracing, as far as possible, the effect of particular patents upon the trade to which they relate, their position with regard to inventions which preceded them and those which followed them, as also the difficulties met with in bringing the idea to a practical issue. Biographical notices have been added in some instances, but the writer regrets that these are very few.

Some of the illustrative matter is absolutely new, and it may here be stated that, as far as possible, the author has avoided repeating the information contained in *Birmingham and the Midland Hardware District*—the standard authority on the trades of the town. For instance, the history of the cloth-covered button has never been given with such fulness before. For this,

# CONTENTS.

the writer has to express his obligations to B. L. Sanders, Esq., of Street Court, Herefordshire, a grandson of the inventor, Benjamin Sanders. The extremely interesting correspondence between Boulton and the Navy Board on the subject of Keir's sheathing alloy—which approaches very closely in composition to Muntz's metal, has not hitherto been published. The writer also believes that he is the first to call attention to Heaton's button-shank machine, as being the precursor of the large family of machines performing a series of distinct operations, all of which are timed by the motion of cams on the driving shaft. Gill's rifling machine, patented as far back as the year 1800, and which contains the elements at least, of the apparatus now in use, seems also to have escaped the notice of former writers.

The author disclaims all intention of having written anything like a formal history of his subject. The most he can hope for is, that the materials here thrown together may be found serviceable hereafter. With this object in view a complete List of Patents granted in Birmingham, from the earliest period down to the passing of the Patent Law Amendment Act in 1852, has been printed in the Appendix, and a very full Index has been given. If in any place the writer should seem to claim too much, it must be attributed to natural partiality for his native place, which is too generally regarded as being exclusively the home of "lacquered shams." He trusts that he has done something to establish a claim on the part of Birmingham, to a high position amongst the centres of mechanical genius.

In conclusion the author has to express his acknowledgments to numerous friends for their advice and assistance, but especially to Mr. SAM. TIMMINS, of Birmingham, for the loan of rare books, letters, and papers, and for his kind help in revising the sheets for press.

H. M. PATENT OFFICE, LONDON,

*July, 1881.*

## CHAPTER XIII.

THE BUTTON TRADE—(*continued*).



## CHAPTER XIV.

THE NAIL TRADE.



## CHAPTER XV.

THE PIN TRADE.



## CHAPTER XVI.

THE NEEDLE TRADE.



## CHAPTER XVII.

THE MANUFACTURE OF METAL TUBES.



## CHAPTER XVIII.

THE MANUFACTURE OF METAL TUBES—(*continued*).

## CHAPTER XXIV.

### Ships' Sheathing and Bolts—*(concluded).*



## CHAPTER XXV.

### Umbrellas and Parasols.



## CHAPTER XXVI.

### The Stamp and Press.



## CHAPTER XXVII.

### Miscellaneous Inventions, 1759-1780.

# BIRMINGHAM INVENTIONS AND INVENTORS.

## CHAPTER I.

IMPORTANCE OF PATENTS IN INDUSTRIAL HISTORY—PATENTS IN BIRMINGHAM AND MANCHESTER — THE FIRST BIRMINGHAM PATENT—BASKERVILLE'S PATENT FOR METAL MOULDINGS ORNAMENTED—BASKERVILLE AS A PRINTER.

DOWN to a very recent period, when the various legal departments were concentrated at the new Law Courts, or, to speak by the card, the Royal Courts of Justice, the entrance to the Patent Office was flanked on one side by the apartments in which the Registrars in Lunacy carried on their operations, and on the other by the rooms devoted to the Registrar in Bankruptcy. That ominous propinquity was explained by some to be typical of the fate which awaited but too many inventors, who only escaped the Scylla of bankruptcy to fall into the Charybdis of insanity. Other cynics have said that the gift of invention, like that of poesy, is but a "fine phrenzy"—in short, a special form of madness. Without entirely adopting this view, which might commit us to the further proposition that every inventor was a fit subject for the issue of a writ *de lunatico inquirendo*, we will admit that there is something to be said for it, and we propose, in the present series of articles, to enquire into the prevalence of the particular complaint in Birmingham and the neighbourhood. It is also our intention to give some biographical details, as far as they can be recovered at this distance of time, of the distinguished persons who have suffered from the malady. To reassure timid readers who have no relish for psychological speculations or medical statistics, let us at once say that we have not the slightest intention of putting their endurance to the test. It is no "philosopher's banquet" that they are invited to, but a repast of humbler and more digestible fare. We simply ask them to take a rapid survey of the various inventions which have made Birmingham not merely "the toyshop of Europe," but a mart for the whole world. Edmund Burke's phrase was all very well at the time, but the town has grown since then, and now aspires to universal empire. Those who may be inclined to regard this as an exaggeration should ask themselves this question: To what part of the globe can you go without finding something which reminds you of Birmingham? It is on record how a body of Arctic explorers who had succeeded in reaching some tremendously high latitude in the frozen North, were startled to find an axe bearing the inscription "Richard Timmins and Sons, Birmingham." Continents are connected by electric cables, manufactured either wholly or in part at Birmingham, and the products of the metropolis of the Midlands are in the hands of the "noble savage" in all parts of the world.

We shall take for our guide the printed specifications of patents, and herein we lay claim to a certain amount of novelty, for, so far as we are aware, it has not hitherto been proposed to write the industrial history of the town by means of the patents granted to residents in Birmingham. It appears to us that these documents have been somewhat neglected as sources of local history, but perhaps the forbidding shape in which they are issued has had something to do with it. No one but an enthusiast reads anything which looks like a "blue book" unless he is obliged; but even professed collectors of topographical miscellanea have passed them over, and we only know one county bibliography in which they appear. We refer to Boase and Courtney's "Bibliotheca Cornubiensis," where every Cornish patent is duly entered under the inventor's name. Charles Lamb would doubtless have included the publications of the Commissioners of Patents in his list of bookss which are no books—*biblia a biblia*—along with "scientific treatises, almanacks, statutes at large; the works of Hume, Gibbon, Robertson, Beattie, Soame Jenyns, and Paley's 'Moral Philosophy.'" But the old specifications have the special value of describing manufacturing processes, often with considerable minuteness, in an age when technical literature can hardly be said to have had any existence. It is not our intention to confine ourselves strictly to those inventions which have been made the subject of letters patent; but, for the reason just given, the information regarding early unpatented inventions is mainly traditionary, and therefore more or less unreliable, whereas the evidence furnished by a specification is unimpeachable, as being that of a solemn legal document.

There is perhaps not another town in the kingdom whose industrial history is so perfectly reflected in the patent lists as our own. This is due to the extent to which Birmingham has always availed herself of the patent laws, and also because of the close connection of the subject-matter of the patents with the various trades of the town. Long ago Bissett "the Magnificent," one of our local bards, in singing the praises of Birmingham, spoke of

> Inventions curious, various kinds of toys,
> Engage the time of women, men, and boys;
> And Royal patents here are found in scores,
> For articles minute—or pond'rous ores.

But it is perhaps known to very few that until the Patent Law Amendment Act of 1852, Birmingham stood first amongst the provincial towns as regards the number of grants of letters patent. The only possible competitor was Manchester (including Salford); but the two combined were absolutely "nowhere" until recent years. Up to the end of the last century 90 patents had been granted to Birmingham, whilst Manchester only reckoned 27. Ten years afterwards—that is, at the end of 1809—they were 122 and 43 respectively, or say nearly 3 to 1. In 1819 the proportion was precisely the same—Birmingham, 187; Manchester, 63. The difference was less striking at the expiration of another decennium, when the figures stood at 268 and 111. At the end of 1839 Birmingham scored a total of 410, against 256 for Manchester; and the latter had gained rapidly when 1849 was reached, the numbers being then 667 and 508. About 1855 they became equal, and at the present moment the balance is in favour of Manchester. In contrasting the growth of the two towns this difference becomes apparent—that whereas the majority of inventions connected with Birmingham trades were indigenous to the soil as it were—that is, they originated in the town itself—Manchester was at first almost entirely dependent upon others. The most remarkable instance of this may be found in the grand invention of "spinning by rollers" (wrongfully claimed by Arkwright), which virtually created a vast industry, giving employment to thousands upon thousands. As we shall show in a subsequent article, Birmingham was the cradle of that invention, which constitutes one of the foundations upon which the wealth and fame of Manchester have been built,

Our treatment of the subject will be, in the main, chronological, but we shall occasionally depart from that order when desirable. For instance, it will be convenient to deal with the inventions of James Watt, which extend over a series of years, in a single article, and a similar course will be pursued in the case of trades whose rise, decline, and fall come within the period embraced in this series. The inventions which originated with the brilliant group who were attracted to Birmingham by the fame of Soho, including Roebuck, Keir, Small, and Murdock, may also be treated as a whole, especially as the subjects of their patents are not very closely connected with the staple trades of the town.

Many of the schemes which we shall have to describe must have turned out failures; but those who have had experience well know that success is not usually attained at a single bound, but rather by a series of slow, laborious, and uncertain steps, which often wander out of the right path. Without entering into minute details, it will be our endeavour to explain the significance of these abortive inventions by a reference to the condition of the particular manufacture at the time. Let no one despise them, for they are not without value. Those who feel inclined to do so should remember that the feeble attempts now being made in untried fields of invention will require considerable indulgence at the hands of posterity. This thought may perhaps induce them to look kindly upon the early struggles of their fellow townsmen. Again, some of the inventions may appear to be in the last degree trivial. So they are. But some of these contrivances have been a source of wealth to their inventors; and when we reflect upon the influence of trifles upon our comfort—we might almost say our happiness—who will say a word against them? But we shall have abundant opportunities of pointing this moral as we go along.

Having placed the subject fairly before our readers, we now address ourselves to the serious business of these articles, and the first patent we have to notice is that granted to Richard Baddeley, ironmonger, on the 22nd of May, 1722, for "A new invention of making streaks for binding cart and wagon wheels, and for smoothing irons of pig iron, never yet before practised in this kingdom, which will be more durable and do three times the service of those made of bar iron, and can be afforded much cheaper." Beyond the fact that this is the first patent granted in Birmingham it has no particular interest, and no details of the invention are given in the official documents. It would appear to consist, firstly, in the application of cast-iron as a material for "streaks," that is the pieces of iron used in place of a continuous tire for binding heavy cart wheels together. An Act of Parliament was passed early in the eighteenth century ordering that cart wheels should be of a certain breadth, and the letters patent sets forth that the invention will be of service in causing the provisions of that Act to be "more punctually observed." As regards that part of the patent relating to irons the grant says "that box smoothing irons being made of the aforesaid metal—that is cast iron—will be of so compact a body as to retain the heat much longer than those made of bar iron, and thereby become more useful and beneficial." It would appear from an advertisement in a local paper in 1746 that Baddeley carried on business in the Square, at the corner of Lichfield Street. The premises were then advertised to be let, and as "further particulars" were to be obtained from Mrs. B., it is to be assumed that Richard Baddeley was no longer alive.

Every lover of the art of printing is familiar with the name of John Baskerville, a versatile genius who deserves a better memorial in the town of his adoption than that afforded by "Baskerville Place," a slum leading out of Broad Street. If our memory is not at fault there was for some years nothing to mark the name of the street until the present inscription in Minton's encaustic tiles was put up about thirty years ago by the late Mr. Richard Prosser, a great admirer of Baskerville, who occupied the premises at the top of the Place. Baskerville was not a native of Birmingham, having been born at Wolverley, in Worcestershire, in 1706. He was originally a stonemason, and about 1736 we find him established in Moor Street as a writing master and a cutter of "gravestones in any of the hands." His gravestones have all perished, but his window-slate is still in existence, and it shows in a marvellous manner the form and style of the "letter," for which he afterwards became famous as a printer. About 1740 he commenced business as a japanner at No. 22, Moor Street; and five years afterwards he took a lease of eight acres of land at "Easy Hill," now Broad Street. In 1742 he took out a patent, the "title" of which is about half a page long, but which admits of being briefly described as a mode of making metal mouldings, and of rolling, grinding, and japanning metal plates. The iron plates are first pickled, and then flatted by being passed through a pair of rolls pressed together by a long lever with a weight on the end. This mode is preferred by the inventor to the ordinary method of employing screws. The plates are next ground by a machine described, and then passed through rolls having the profile of the intended moulding cut in them. The mouldings being japanned or varnished, are intended to be applied in making picture-frames and ornamenting furniture of all kinds. The inventor also proposed to japan or varnish the flat plates, which he uses for any objects "now usually veneered with ebony, whalebone, walnut, mahogany, peartree, or otherwise. The said plates so prepared and japanned or varnished being more beautiful and durable, and in all respects answer better, than anything which has hitherto been applied to the same purpose, as the same will produce a fine glowing mahogany colour, a black in no way inferior to that of the most perfect India goods, or an imitation of tortoiseshell which greatly excels nature itself (!), both in colour and hardness, and each colour admitting of the most perfect polish, whose beauty, without violence, will not be impaired in several ages." This is the first patent for making metal mouldings by passing them through rolls of a certain profile, though the idea was patented several times subsequently. Later on we shall find the draw-bench used for the same purpose. No improvement is claimed or described in the particular process of japanning.

Had he stopped here his name would probably never have been known outside the town in which he lived: but in 1750 he turned his attention to printing, and in 1757 he issued a noble quarto volume of Virgil's Æneid, which has been highly prized by all lovers of the printing art. Six hundred pounds, it is said, were expended before he could produce a "letter" to please his fastidious taste. Not only did he design his own type, but he was his own paper-maker; he compounded his own ink, and made his own presses. The results are before the world in the many books which he printed; and in our humble judgment the works of the Birmingham japanner are not unworthy to be placed by the side of the acknowledged masterpieces of those whose whole lives were devoted to the typographic art. From a characteristic letter which he addressed to Horace Walpole in 1762, it appears that the printing business was not a commercial success. Baskerville expresses himself as being "heartily tired of it," and repents ever having attempted it. "It is surely (he says) a particular hardship that I should not get bread in my own country (and it is too late to go abroad) after excelling in the most useful art known to mankind; while every one who excels as a player, fiddler, or dancer lives in affluence, and has it in their power to save a fortune." He died January 18, 1775, and was buried in his own garden, one of his peculiarities being a strong antipathy to interment in consecrated ground. There his body remained until about 1825, when it was dug up during the excavations for the canal. Some uncertainty prevailed as to the final resting-place of Baskerville's body until quite recently, when it was satisfactorily ascertained that it was removed to Mr. Knott's vault at

Christ Church, where it now rests. Some very interesting communications on this subject appeared in our "Local Notes and Queries" during the autumn of last year.

## CHAPTER II.

THE INVENTION OF SPINNING BY ROLLERS—JOHN WYATT AND LEWIS PAUL—THE BEGINNING OF COTTON SPINNING IN BIRMINGHAM — ARKWRIGHT'S LATER PATENTS.

ALLUSION was made in the introductory article to the fact that Birmingham was the cradle of one of the inventions which have made Manchester famous. The tale is a sad one, and is full of disappointment, privation, and ruin to most of the actors. The principal character is one Lewis Paul, about whose parentage there is some doubt. Mr. Smiles says in his "Huguenots" that he "was the son of a French refugee, who carried on business as a druggist in St. Paul's Churchyard," but his name does not occur in the roll of the Apothecaries' Company, and it is not likely that any unlicensed person would have been allowed to pursue his calling only a hundred yards or so from the headquarters of a society which has always been most jealous of its privileges. Be this as it may, the father seems to have accumulated considerable property, and on his death young Lewis was left under the guardianship of Lord Shaftesbury and his brother, the Hon. Maurice Ashley Cooper. The details of the early life of this remarkable man are not extant, but we know for certain that he was the inventor of a machine for pinking crapes, tammies, &c., for shrouds, which brought him considerable profit. Mrs. Desmoulins, Dr. Johnson's protegée, frequently referred to in Boswell, was employed in working the machine. Pursuing the career which he had chosen for himself, probably from necessity, as he seemed to have run through his patrimony when a young man, we find that he associated himself with one John Wyatt, who is described in a deed of agreement, dated 1732, and having reference to a file-cutting machine invented by the latter, as "of the parish of Weeford, county of Stafford, carpenter." Wyatt was born in 1700, but whether he was a native of Weeford, a small village about four miles from Lichfield, we have not been able to ascertain. It has been asserted that he was a Birmingham man, but perhaps on insufficient authority. Paul was then residing near Holborn, and the exact date of his removal to Birmingham is unknown, but it must have been at some period prior to 1738, the date of his first patent, in which he is described as of "Birmingham, gentleman." The title of the invention is "A new invented machine or engine for the spinning of wool and cotton."

The machines employed in the preparation of textile fibres are usually of an exceedingly complicated character, at all events in appearance; but when stripped of extraneous details the principle, in the majority of cases at all events, is found to be extremely simple. The object of Paul's machine was to stretch the "slivers" of wool or cotton, and thus bring it one step nearer to the condition of yarn or thread. Perhaps we should explain that a "sliver" is a loosely assembled untwisted mass of fibres laid side by side. It has no strength or tenacity, and when pulled gradually becomes finer and finer, until at length it comes apart. Now, "spinning by rollers," as Paul's invention is technically termed, is nothing more than stretching slivers in a regulated manner, and it is accomplished by mechanism which is exquisite in its simplicity, consisting merely of two pairs of rollers, placed a few inches apart, through both of which the sliver passes. If both pairs are moved with the same velocity the sliver will come out unaltered in length, cotton fibres not being malleable or capable of extension by mere pressure; but increase the velocity of the second pair, and what happens? Suppose, for example, that the first pair revolves with a velocity of 12 inches per second, and the other with a velocity of 18 inches in the same time. The grip being too firm to permit slipping, the velocity of the sliver must be the same as that of the rollers, so that a sliver which measured 12 inches when it went in measures 18 inches when it comes out. In other words, it has been stretched or "drawn" until it is half as long again; and the amount of this "drawing" may obviously be regulated to the greatest nicety by varying the rate of speed of the second pair of rollers. It is an exquisite delight to listen to

> Notes with many a winding bout
> Of linked sweetness long drawn out.

but the mechanician feels a pleasure all his own in contemplating a perfectly designed machine such as this is.

To trace the history of the invention at the length which its intrinsic importance thoroughly deserves would require at least thrice the space which we can allot to this particular phase of the industrial history of the town. Amongst the losses which we have to deplore by the disastrous fire at the Free Library is that of a very large collection of original documents and papers brought together by the late Mr. Robert Cole, having reference to the long-continued endeavours of Wyatt and Paul to bring their invention to a successful issue. They succeeded in 1741 in setting up a mill in a warehouse near the Well, in the Upper Priory, exactly on the site of the present Friends' School Rooms; the machine being worked "by two asses walking round an axis." A piece of the thread spun at that time is still extant, endorsed with the date in Wyatt's own handwriting. The desperate expedients to which the inventors were reduced are exceedingly painful to read. Wyatt was frequently compelled to pawn his clothes, and to deny himself almost the necessaries of life, to obtain the means for perfecting a machine which subsequently became a source of affluence to entire districts. Things began to look a little brighter when a well-known Birmingham bookseller, Thomas Warren by name, came to their assistance with an advance of £1,000., but the concern did not prosper. Warren became a bankrupt, and in 1742 Wyatt was a prisoner for debt in the Fleet. The invention seems to have possessed an irresistible attraction, and many celebrated men were mixed up in the speculation. Dr. Johnson, Dr. James—a physician who once practised in the town, and the inventor of the well-known fever powders which bear his name—Edward Cave, founder of the *Gentleman's Magazine*—a Warwickshire man, by the way—and others of lesser note. We may also mention Samuel Garbett, described in the destroyed papers as a "ringmaker." We take this to be the same Garbett who subsequently became Roebuck's partner in founding the sulphuric acid trade in this country; but more anon on this point. Later on an attempt was made to revive the manufacture in Northampton, but it failed there as it failed here. The project appears to have excited a considerable amount of attention, and it is thus described by the Rev. John Dyer in his poem of "The Fleece," which appeared in 1757:—

"A circular machine, of new design
In conic shape: it draws and spins a thread
Without the tedious toil of needless hands.
A wheel invisible, beneath the floor,
To ev'ry member of th' harmonious frame,
Gives necessary motion. One intent
O'erlooks the work; the carded wool, he says,
So smoothly lapped around those cylinders,
Which gently turning, yield it to yon cirque
Of upright spindles, which with rapid whirl
Spin out in long extent an even twine."

As poetry, this is sad stuff, but it could hardly have been written by one who had not seen the machine in actual operation, and is, so far, very valuable evidence.

It was remarked in the opening article that the invention of spinning by rollers was claimed by Arkwright, and it is, in fact, embodied in his patents of 1769 and 1775. It is on record that he was aware of the labours of his predecessors, for he refers to them in his "Case," laid before the House of Commons in 1782, and it is believed by some that he was well acquainted with Paul and Wyatt's invention from the beginning of his career. That he wished to descend to posterity as the inventor of spinning by rollers is evident from the fact that he caused the spinning rollers to be introduced as the only adjunct in the well-known portrait of him painted by James Wright. Some discussion has arisen with regard to the share which Wyatt had in the invention, and though he was undoubtedly of great assistance to Paul, some are inclined to attribute the whole of the credit to him. This is, however, a point on which we need not enter, more especially since Paul's claims were ably set forth some years ago in a paper by Mr. Robert Cole, which is printed as an appendix to French's "Life of Crompton." It is certain that the invention took its rise in the town, and that is sufficient for our present purpose.

In spite of his failure, Paul still went on inventing until the close of his life, and in 1748 he obtained a second patent for "a machine for carding of wool and cotton, &c.," which contains an essential feature of the carding engine of the present day. Carding by hand was performed by spreading the fibre on a board furnished with a series of projecting teeth. Another similarly furnished piece of board, held in the hand, was used to comb the cotton thus placed between them, until all the fibres were laid parallel. Paul proposed to place these teeth on a cylinder, which was caused to revolve very nearly in contact with a concave frame lined internally with a similar set of teeth, the cotton being fed in between the two surfaces. There was a contrivance for letting down the concave part and turning it round, so as to permit of the cotton being stripped off. This was done by "a stick with needles in it parallel to one another like the teeth of a comb." The length of the cardings depended, of course, upon the circumference of the carding cylinder; but an ingenious apparatus was attached for making them into a continuous carding. This bears the closest resemblance to the modern carding cylinder, except that the concave frame is now placed over the cylinder instead of under, as in Paul's machine. "Here then," says Mr. Baines in his "History of the Cotton Manufacture," "are the carding cylinders, the perpetual carding, and the comb for stripping off the carding. It must be admitted that the invention was admirable and beautiful, though not perfect. Its defects were that the cylinder had no feeder, the wool being put on by the hand: that the cardings were taken off separately by a movable comb, which, of course, required the machine to stop; and that the perpetual carding was produced by joining short lengths with the hand; whereas now it is brought off the machine in a continuous roll by a comb attached to the cylinder, and constantly worked against it by a crank." Paul does not seem to have derived any benefit whatever from this ingenious invention, and on the breaking up of the establishment at Northampton the machine was purchased by a hat manufacturer from Leominster, and by him applied to the carding of wool for hats. About 1760 it was introduced into Lancashire, and reapplied to the carding of cotton by a gentleman of the name of Morris, in the neighbourhood of Wigan. Baines goes on to say that Mr. Peel, of Blackburn, was one of the first to adopt it. This Mr. Peel was the great-grandfather of the present Sir Robert, and was the founder of the family. He was familiarly known as "Parsley Peel," from the pattern of a printed calico which he introduced.

In 1758 Paul took out another patent for spinning, which, though very ingenious, does not call for any particular remark here. He had then returned to London, and he is described in the specification as of Kensington Gravel Pits. His death took place at Brook Green, in April, 1759, and he was buried on the 30th of that month in Paddington Churchyard.

Before bringing the history of this remarkable episode to a conclusion, it will be well to glance briefly at the fate of Paul's fellow-labourer. After the final break up of the spinning concern, Wyatt seems to have found employment as a workman under Boulton. He died November 29, 1766, aged sixty-six, and was buried in St. Philip's Churchyard, in the triangular plot opposite to the Rectory and the Bluecoat School, about midway between them, and near the railing. To show the esteem in which he was held, Boulton himself attended the funeral, and so did Baskerville—"the latter in a splendid suit of gold lace!" The gravestone was restored a few years ago by Mr. John Rabone. Wyatt is credited with many ingenious inventions, but no particulars have been preserved of any of them except the spinning machine. Amongst them was a machine for cutting files, a weighing machine, one of which was set up in Birmingham a few years before his death.* His own description of it, as given in a letter by his son in the "Repertory of Arts" for January, 1818, was that "it would weigh a load of coal or a pound of butter with equal facility, and with nearly equal accuracy." His name is also traditionally associated with certain improvements in the lathe and tools used for making pearl and bone buttons. There is documentary evidence for the fact that he invented a file-cutting machine, but we are not aware that there is any authority beyond tradition for the statements as to his other inventions. A list of them was given in an article which appeared in the *Mail* for February 27, 1877.

## CHAPTER III.

BOULTON AND THE SOHO MINT — THE HYDRAULIC RAM—DR. ROEBUCK AND THE CHEMICAL TRADES—MANUFACTURE OF SULPHURIC ACID.

### BOULTON AND THE SOHO GROUP.

THE praises of Boulton and Soho have been sounded so often and with so much ability by others, that we must decline the tempting opportunity which offers itself to tell once more the story of that brilliant episode in the history of the town. We have only to deal with Boulton and the group of men by whom he was surrounded as inventors, and to relate in a matter-of-fact way how they contributed to the department of applied science. The men who were gathered round him were men of widely different temperaments, and of various acquirements, and we are conscious of a certain anachronism in associating Boulton, Roebuck, Small, Keir, Murdock, and Southern. But, as we have remarked in a previous article, this has been done solely as a matter of convenience. All were eminent in their particular line, and some would have attained excellence in almost any career which they might have chosen.

### MATTHEW BOULTON, F.R.S.

We begin then with Matthew Boulton, who was born in the town on the 3rd September, 1728—his father, also

* It appears from an official announcement, quoted in Mr. Langford's "Century of Birmingham Life," that a new "Town machine" was erected in the early part of the year 1752. Whether this was Wyatt's machine we are unable to say.

named Matthew, having been a native of Lichfield, but belonging to an old Northamptonshire family. He received the rudiments of his education at a school kept by the Rev. Mr. Ansted in Deritend, and at an early age he commenced work with his father, who carried on business as a toy maker. Mr. Smiles says that when he was only seventeen "he introduced several important improvements in the manufacture of buttons, watch chains, and other trinkets; and he had invented the inlaid steel buckles which shortly afterwards became the fashion." We are unable to say precisely what these improvements consisted in, and we may remark, generally, that Boulton's place in this series will be apparently insignificant by reason of the meagreness of the records which have come down to us. Though it may sound like a paradox, it is perfectly true to say that if he had not been so great a man we should have known more about him as an inventor. The fact is, that his labours as the directing mind of a gigantic manufacturing establishment, and the introducer of a machine which has totally revolutionised our habits of life and our mode of thought have overshadowed his mere inventions. That he possessed a true mechanical instinct is certain from what has come down to us, and we have the testimony of James Watt himself to the same effect, though it is not likely that he had time to work out all the details of his improvements. It appears from memoranda found amongst his papers, that he was at one time engaged in trying to improve the guns of the period, by greater attention to the accuracy of the bore, and by the use of a telescopic sight. He also suggested cylindrical shot with a parabolic end, as offering less resistance to the air. When, however, the partnership with Watt was commenced, he was for years immersed in the commercial part of the undertaking, the difficulties of which were so enormous that they almost overcame even his indomitable, sanguine, and buoyant spirit.

Perhaps Boulton's mechanical genius was most conspicuously displayed in his improvements in coining, and the productions of the Soho Mint had at one time a world-wide reputation. He was engaged for many years in trying to rouse public opinion to the disgraceful state of the national coinage, and at length he succeeded. It was bad artistically, and the mode in which it was produced was equally faulty. In 1790 he took out a patent for an improved mode of working coining presses, which was also applicable to presses for stamping or cutting out. The presses were arranged in a circle round a large horizontal wheel which was revolved continuously by the prime mover —steam engine, water-wheel, windmill, or what not. On this wheel a series of cams, or curved projections, were placed, which came successively into contact with certain mechanism by which the screw of each press was raised. The power required for delivering the blow was obtained by atmospheric pressure, and each press was furnished with a cylinder closed at the bottom, but open at the top, and fitted with an air-tight piston. At the same time that the main wheel set the mechanism in operation for raising the die, it also drew up the piston in the closed cylinder, thus creating a vacuum in the lower part. The piston had of course a tendency to descend, owing to the pressure of the atmosphere, and this was applied to the fly of the press by means of a chain connected with the piston, and passing over a pulley. The various motions were so timed as to release the connections which allowed the cylinder to descend at the right moment, and thus bring the die down with a powerful blow upon the coin to be struck. Other modifications are described in the specification, according to one of which the air-cylinders may be replaced by the cylinders of a steam engine, in which the vacuum is to be produced in the ordinary way, or an exhausting pump may be employed. The leading idea is, however, the use of atmospheric pressure to give the blow.

It was not until 1797 that the Government entrusted a contract for copper coinage to Soho, and shortly before the close of Boulton's career he was engaged in fitting up the new Royal Mint on Tower Hill, with machinery made according to his patent. It was set to work in 1810, and it will probably be a surprise to many to be told that some of the original machinery is at work at this very hour. It would be impossible to produce stronger evidence than this of the excellent workmanship and design which characterised the productions of Soho in its palmy days.

In 1797 a patent was granted to Boulton for "an apparatus for raising water and other fluids,' which Mr. Smile, naïvely describes as a mode of "raising water by impulse, somewhat after the manner of Montgolfier's hydraulic ram." The fact is, the invention *is* the hydraulic ram, neither more nor less, and Boulton was not the inventor at all, he simply acted as agent in the matter at Montgolfier's request. There is nothing whatever in the specification to show that the invention was "a communication," as it would now be called, and this omission has given rise to a good deal of mystification, and has perhaps caused some angry *réclamation* by the friends of the distinguished aëronaut. Not however that he was the inventor of the hydraulic ram, which is due to an Englishman named Whitehurst.

Boulton died on the 17th of August, 1809, aged 81, and was buried in Handsworth Church, whither he was followed by six hundred Soho workmen.

### DR. JOHN ROEBUCK, F.R.S.

We have now to speak of a man who, like Boulton, was bold and prompt in action, of a keen and active spirit, but who lacked the prudence which led his friend to fame and fortune. Had Roebuck possessed but a small portion of what is roughly called "business capacity" he might have achieved a success little inferior in its way to that of the founder of Soho.

John Roebuck was a native of Sheffield, where he was born in 1718. His father was a cutler in the town, and is was intended that he should follow the same business, but a strongly developed tendency towards scientific pursuits induced the father to allow his son's wishes to prevail. After having studied at the University of Edinburgh, he proceeded to Leyden, where he graduated in medicine. In 1745 he settled as a physician in Birmingham, an opening having presented itself by the death of a practitioner in the town. His skill and engaging manners brought him a good practice, but his ruling passion was experimenta- chemistry, and much of his time was spent in his laboratory. He seems indeed to have been what what we should now call a "consulting chemist," and it is on record that the local manufacturers were in the habit of resorting to him whenever any chemical difficulty occurred. The laboratory was carried on by his partner, S. Garbett, until the close of the century. Indeed Birmingham seems to have very early appreciated the importance, from a manufacturing point of view, of the growing science of chemistry, and a work was published in the town in 1790 on "The Chemical Principles of the Metallic Arts: designed chiefly for the use of Manufacturers," by W. Richardson, a local medical practitioner. In an appendix, the author deals with the diseases to which the workmen engaged in various trades are more particularly liable. To return, however, to Roebuck. His mind was eminently practical, and his investigations soon began to bear fruit. While he was yet a young man of 25 he had actually laid the foundation of what was practically a new industry, which has since his time attained enormous dimensions, not only in the town itself, but in many other parts of England. We refer to the manufacture of sulphuric acid, of which Liebig remarks in one of his inimitable "Letters on Chemistry," that "it is no exaggeration to say, we may judge with great accuracy of the commercial prosperity of a country from the amount of sulphuric acid it consumes." As most of our readers are aware, sulphuric acid, or oil of vitriol, to use the familiar synonyme, is a most powerful solvent. There were at one time considerable difficulties in finding a material suitable for making vessels which would withstand its action during the process of manufacture. This, by the way, was a difficulty which does not seem to have troubled the old alchemists, who spent so much of their time in seeking for the "alkahest," or universal solvent. Before Roebuck's day glass vessels were employed. These were necessarily of small size, and very liable to break; but Roebuck hit upon the happy idea of using chambers made of sheet lead, and, acccording to Parkes's "Chemical Essays," a manufactory was started in 1746, which very soon had the effect of reducing the price of the acid to one-fourth of its former cost. It is quite true, as Dr. Percy points out in his recently-published "Metallurgy of Silver and Gold," that Glauber, in the 17th century, had used leaden vessels for condensing acid fumes, but still Roebuck must be credited with the first practical application of the fact that lead would resist the action of sulphuric acid, and of establishing "a new manufacture within the realm," to quote the words of the Statute of Monopolies. In conjunction with his partner, Samuel Garbett, whom we have already mentioned in connection with Paul's cotton-spinning concern, the manufactory was carried on for many years, and was a source of considerable profit. The works were in Steelhouse Lane, and they eventually passed into the hands of Messrs. Alston and Armitage, and were continued under the name of Alston and Sons down to the year 1825, and perhaps later. Alston, we should remark, is a well-known name in Birmingham. The process was never patented in this country, but Roebuck and Garbett obtained a patent for Scotland in 1771. By that time the method had become pretty generally known, although attempts were made to keep it secret, and the patent was pronounced to be bad for want of novelty by the Scotch Courts, which decision was upheld by the House of Lords in 1774. The patentees had themselves established an oil of vitriol works at Prestonpans in 1749; and according to Parkes, the manufacture was commenced at Bridgnorth in 1756 by a workman who had absconded from Roebuck's employ at Birmingham.

The success of the Prestonpans works seems to have drawn the doctor's attention to Scotland as a field for his restless, enterprising spirit. It would be foreign to our purpose to follow him in his career as the founder of the Carron ironworks, by which he became fully entitled to the credit of being the father of the Scotch iron trade. He relinquished his practice as a physician at a date which we have not been able to ascertain, and spent much of his time in Scotland. In 1762 he took out a patent—in which he is described as residing in Birmingham—for making malleable iron from pig-iron by the use of raw coal instead of charcoal, as was then the practice. The specification is very meagre, and consists of about half a dozen lines only. The process could only have been a success in the event of his being able to prevent the fuel from coming into direct contact with the metal, and only use the *flame* of the coal.

Shortly after this date we find him in communication with the Government on the subject of certain attempts then being extensively made to entice workmen abroad—a proceeding which was at that time contrary to the law, and was viewed as a high crime and misdemeanour. There s a letter of his to the Secretary of State, dated June 2, 1764, preserved amongst the Home Office Papers at the Public Record Office, in which he encloses copies of documents contained in a pocket-book lost by a passenger in the coach from Birmingham to London. The pocket-book had been handed to Roebuck, under the supposition that he might possibly be acquainted with the owner, one of the papers being a certificate of character, signed by Roebuck and his partners Garbett and Cadell, given to Lloyd, one of their workmen. Lloyd and several others it appears had arranged to go over to Sweden to erect an iron furnace, and the pocket-book contained various agreements stating the terms upon which they were to be employed Roebuck's zeal was "much commended," and the introduction led to a long correspondence. A little later on Garbett urges the Government to take steps to prevent the fraudulent importation of iron in bars and plates as "unmanufactured" iron, which paid a lower rate of duty. As there appeared to be some doubt about the matter, the authorities were unwilling to contest the matter unless Roebuck and Garbett would guarantee the cost of a trial in the Exchequer. This they agreed to do, as it was obviously their interest, as ironmasters, to throw difficulties in the way of the importation of iron from abroad. The Government were so pleased with Roebuck's conduct that Lord Halifax, Secretary of State, was induced to trust him with a mission "of a very secret nature, in order that he might obtain information." The Government had received intelligence that "one Boulton" was engaged in a scheme for conveying British workmen to Sweden, and Roebuck was requested to watch Boulton's movements, and report to Lord Halifax ! Roebuck was then absent in Scotland, and the letter was replied to by his secretary, Farquharson, who we think subsequently went to Soho, to the effect that Boulton was a man of very high character, and not in the least likely to lend himself to any such transaction. The letters are very amusing. Garbett appears eventually to have become on very friendly terms with the officials at the Treasury, and in a letter dated May 4, 1766, we find him inviting Edmund Burke to come and spend a few days with him in Birmingham, adding that Mrs. Garbett and family would be in Scotland until Christmas, so that "he will have quite a bachelor's house."

According to an old Directory of the town, Garbett in 1770 was carrying on business as a merchant at No. 11, Newhall Street, and he seems to have played no inconsiderable part in local affairs for many years. In his later days he was chairman of the "Commercial Committee"--a sort of precursor of the Chamber of Commerce of our times, and in 1790 a warm dispute arose as to the manner in which he had expended the funds of the Committee in opposing the bill for permitting the exportation of brass. As might have been expected, Matthew Boulton took part in the discussion, and the whole story may be read in Mr. Langford's book.

The final overthrow of Roebuck's schemes, and his ultimate ruin—which was a source of deep grief to Watt and Boulton—are well told in Mr. Smiles's "Lives of Boulton and Watt," and also in a special chapter of that author's "Industrial Biography." A detailed memoir of his life and works was read before the Royal Society of Edinburgh in 1796. How he became connected with Watt, to whom he was of material assistance in his early career as an inventor, we shall relate in a future article. Roebuck died July 17, 1794. It is, perhaps, not generally known that the late Mr. J. A. Roebuck, who represented Sheffield in the House of Commons for so many years, was his grandson.

## CHAPTER IV.

THE SOHO GROUP — WILLIAM SMALL — HOROLOGY— ISOCHRONOUS PENDULUM—CHEMICAL MANUFACTURES — COMPOUND METAL — SHIP SHEATHING — WHITE LEAD.

### DR. SMALL.

William Small was born in 1734 at Carmylie, in the county of Angus, where his father was minister. He was appointed Professor of Natural Philosophy at the University of Williamsburg, Virginia, but the climate not suiting his health he returned home, and settled in Birmingham. In 1765 he commenced practice as a physician at No. 9, Temple Row, and he appears to have acquired considerable eminence in his profession. He was introduced to Boulton by the celebrated Benjamin Franklin, and was the intimate friend and close correspondent of Watt, with whom he became acquainted during a journey to England made by the latter in 1768. The pair were in the habit of exchanging ideas upon the subjects which happened to be engaging their attention at the time, and many of the letters are printed at length in Muirhead's "Life of Watt." Small was a man of singular amiability of character, and he appears to have been a sort of connecting link between the various members of the Soho group. He was largely instrumental in bringing Boulton and Watt together, and was in all probability the person who actually introduced them on the occasion of their first meeting. Small possessed a truly philosophic intellect, and busied himself, as his letters to Watt fully show, with many scientific problems. Horology was one of them, and he took out a patent for improvements in clocks in July, 1773. The specification includes several distinct inventions, the gradual perfection of some of which may be clearly traced in his letters to Watt. One of his notions was to dispense with hands, and to mark the time by means of a revolving graduated dial, the smaller intervals being read by means of a vernier. In another part of his specification he describes a mode of rendering the vibrations of a pendulum isochronous by using a particular form of bearing. He also proposed to govern the motion of clocks by immersing the whole or part of the mechanism in a liquid, an idea which has been lately revived in Durham's governor. Several other contrivances are also described in the specification, but they need not be further alluded to here. Mention is made in some of the Soho papers of "a clock with one wheel" of Small's invention, but there is nothing of the kind in his patent.

Small's death took place in 1775, at the early age of 41, just at the time when it was contemplated to take him into partnership at Soho. Boulton felt his loss most keenly, and judging from the tenour of a letter which he wrote to Watt at the time, an attachment subsisted between the two of unusual depth and force. Watt himself was deeply affected, and the manner in which the news of his death was received by Keir, Darwin, and John Day, the author of "Sandford and Merton," shows that Small was no common man. Keir speaks of him as one "who, to the most extensive, various, and accurate knowledge in the sciences, in literature, and in life, joined engaging manners, a most exact conduct, a liberality of sentiment and enlightened humanity."

## JAMES KEIR.

The subject of this notice was, perhaps, the most brilliant and versatile of the Soho group. Doctor, soldier, manufacturer, chemist, and philosopher, he seems to have been "everything by turns and nothing long." He belonged to an ancient Scottish family, and was born at Edinburgh September 29, 1735. Whilst studying medicine at the University of his native city he became acquainted with Dr. Darwin, and the friendship thus formed was never interrupted. Abandoning his intention of following the profession of medicine, Keir entered the army, and served with the 61st Regiment in the West Indies and in Ireland. He quitted the army about 1770 to settle in Birmingham, and a few years later we find him established as a glass maker at Stourbridge, but he abandoned the business in 1778. Two years afterwards he started works at Tipton on a very extensive scale for the manufacture of alkali, soap, white and red lead, and metal window sashes. Shaw, in his "History of Staffordshire" (ii. 136), states that the works occupied "a site of several acres adjoining the Birmingham canal," adding that "there is not a single person, from the proprietors downwards, who are now employed in these manufactures who had been bred to them."

Keir took out two patents, the first being dated December 10, 1779, for "A compound metal capable of being forged when red hot or when cold, more fit for the making of bolts, nails, and sheathing for ships than any metals heretofore used or applied for those purposes." This is a very important patent, and has given rise to much discussion, as it has been held to contain a clear anticipation of the well-known "Muntz's metal." Keir says: "My compound metal consists of certain proportions of copper, zinc, or spelter, and iron, combined together by means of heat or otherwise—that is to say, of 100 parts of copper, 75 parts of zinc or spelter, and 10 parts of iron. When those proportions are well combined together, they constitute a compound metal, which may be forged either when cold or when heated to a blood red. Although these proportions are the best, and give the greatest strength and malleability when hot and when cold, yet other proportions, in which the iron, or the zinc, or the spelter may be respectively diminished or increased within certain limits relatively to the copper, will be found to possess an inferior degree of these properties." The remainder of the specification is occupied with the precautions to be observed in melting and mixing the metals together.

The similarity of this composition to Muntz's metal was, we believe, first pointed out in "Birmingham and the Midland Hardware District," a series of reports on the trades of the town, published at the time of the visit of the British Association in 1866. This gave rise to some remarks by the late Matthew Davenport Hill, Q.C. (in support of Mr. Muntz's claims), in the course of his address to the Midland Institute in September, 1867, and a long discussion followed in our "Local Notes and Queries" column. Mr. Hill was thoroughly acquainted with the subject, for he was Mr. Muntz's counsel in several actions brought by that gentleman against the infringers of his patent. The compositions of three very similar alloys which have been proposed for ships' sheathing are as follow :—

| | Keir—1779. | Collins—1800 | Muntz—1832 |
|---|---|---|---|
| Copper | 54 | 55·5 | 50—63 (60) |
| Zinc | 40.5 | 44·5 | 50—37 (40) |
| Iron | 5·5 | — | — |

Mr. Muntz's claim is a wide one, embracing as it does all alloys between the limits above given, a composition consisting of 60 parts copper and 40 zinc being however preferred. The patent of Collins was brought forward against Muntz, in one or two trials for infringement, but the judges held that it was not an anticipation. Collins was, we believe, at one time a Birmingham button maker, though he seems to have removed to London at the latter end of the last century. In 1790, one William Collins, of Lambeth, had a patent for making pipes and other articles out of tinned copper, in conjunction with Charles Wyatt of Birmingham, whom we verily believe was the son of the John Wyatt who was associated with Lewis Paul in the cotton-spinning experiments in Birmingham, as described in our second article.

We have no desire to deprive the late Mr. Muntz of the credit to which he is undoubtedly entitled for the successful introduction of the metal bearing his name, and which eventually yielded a splendid fortune; nor do we for a moment believe that he was not an independent inventor. Our point is rather to show that Keir's scientific knowledge enabled him to compound a metal which would roll when hot, and would also possess the property of slowly oxidizing, and so keep itself free from seaweed and barnacles. During the discussion above alluded to, a point was attempted to be made by showing that no sheathing was ever made under Keir's patent : but that is a great mistake. In the autumn of 1779, some months before the grant of the patent, Matthew Boulton brought the matter under the notice of the Admiralty, in a letter to the Earl of Sandwich, and he solicited that the Government would order a trial to be made of the new sheathing. Boulton stated that the composition was the joint invention of himself and Keir, that it was harder than copper and less liable to oxidise, and that it was very suitable for ships' bolts, as it stood the blows of the mallet better than copper when being driven. The Board appear to have thought favourably of the application, and the Juno frigate, then in course of construction, was placed at the inventors' disposal. A sufficient quantity of sheathing was got ready, and Keir was ordered to attend at the Admiralty to make the necessary arrangements, but whether he ever did so we are unable to say. These facts are now published for the first time, and they furnish indisputable evidence that Keir's metal was not a mere experiment. It appears further from Shaw's account of the works at Tipton, quoted above, that window sashes were made "of a peculiar golden-coloured metal—a compound of copper, zinc, and iron—invented by the proprietors, and possesses singular properties which make it fit for this purpose. It is as strong as iron, and, like iron, may be forged or rolled when red hot. Bars of this metal are rolled hot between steel rollers, in which concavities are so cut and turned that they give corresponding convex mouldings to the bars as they pass through them. Windows of this manufacture are in Windsor Castle, Carlton House, and in many principal houses in the kingdom."

Keir also took out out a patent for the manufacture of white lead in 1806, but he appears to have used a certain "new process" many years before. Shaw, writing in 1801, says that it "differs from the common mode, in this respect, that vinegar is not employed." The method described in the specification is the precipitation of the carbonate from a solution of chloride of lead by the addition of an alkaline carbonate. It is not of any special industrial importance.

Besides being a manufacturer, Keir was a man of profound chemical knowledge, and but for a certain amount of restlessness, as is shown by his numerous pursuits, he would doubtless have achieved a considerable reputation. He contributed several memoirs to the "Transactions" of the Royal Society, of which he was a Fellow, amongst which may be mentioned papers on the crystallisation of glass (1776), the freezing of sulphuric acid (1787), and especially one on "the dissolution of metals in acid, and their precipitation, with an account of a new compound acid menstruum, useful in some technical operations of parting metals" (1790.) The menstruum referred to is what is known as " stripping liquor," and is composed of a solution of nitre in sulphuric acid. It is used for obtaining silver from scrap plated metal, and has the peculiar property of dissolving silver, while it does not attack copper. He also published a "Treatise on the various kinds of permanently elastic fluids or gases" (1777); a translation of Macquer's "Dictionnaire de Chimie," to which he added some valuable notes; a "Dictionary of Chemistry," which was never completed, and a Life of his friend Day, the author of "Sandford and Merton."

Keir was intimately connected with Boulton, Watt, Priestley, Withering, and the other distinguished men who had made Birmingham their home, and he was in close correspondence with Berthollet, De la Metherie, and the great French chemists of the day. At one time it was contemplated to take him into partnership at Soho, and he took temporary charge of the concern during the absence of the heads of the firm. Eventually he became a partner with Watt in the copying-machine business. He died October 11, 1820, and is buried in West Bromwich churchyard. His great grandson, Mr. J. Keir Moilliet, has published an interesting memoir of his distinguished ancestor,

## CHAPTER V.

WILLIAM MURDOCK: HIS INVENTIONS — ARSENICAL PAINTS—GAS LIGHTING—ROTATORY ENGINE—SLIDE VALVE—BORING MILL — STONE-PIPES—PNEUMATIC BELLS AND DISPATCHES—IRON CEMENT—SUN AND PLANET MOTION--JOHN SOUTHERN AND BALLOONS.

### WILLIAM MURDOCK.

The ablest lieutenant the Soho firm ever had was unquestionably William Murdock, a Scotchman by birth, the son of a millwright and miller of Bellow Mill, near Old Cumnock, Ayrshire, where he was born in 1754. Little or nothing seems to be known of his early life until 1777, when he presented himself to Boulton, at Soho, and asked for a job. The story goes that Boulton's attention was attracted by a curious hat which the young man had—a *wooden* hat, be it known—turned "on a wee bit lathey of my own making," as Murdock said in answer to his future master's enquiry.

On the strength of this story, it has been asserted that Murdock was the inventor of the oval lathe; but we may say that the machine was well known and was described in a book before he was born. Murdock was at once taken on, and was shortly afterwards engaged for two years at a salary of 15s. per week when at home, 17s. when out, and 18s. when in London. It appears to have been the custom at Soho to bind the workmen for long periods, no doubt to prevent them from being enticed to other employment. The business of the firm in Cornwall was so extensive that it was found necessary to have a permanent representative on the spot, and Murdock took up his residence at Redruth, where he remained until 1798 or 1799. In 1784 his attention was directed to the subject of propelling carriages by steam, and he made a small model locomotive, which, though not the first of its kind, was a remarkable novelty. Cugnot was at work on the subject in France as early as 1769-70, and his road-locomotive is still preserved in the Conservatoire des Arts et Métiers at Paris. A model to scale was shown at the Loan Collection of Scientific Instruments at South-Kensington in 1876. When Murdock's locomotive was finished he proceeded to try it in the long avenue leading to the parsonage at Redruth, and the story goes that the worthy pastor was nearly frightened out of his wits by meeting the hissing fiery monster during his evening walk. The model is still in existence, and is in the possession of the family. It has been publicly exhibited many times, and in 1850 it was shown at work at a meeting of the Institution of Mechanical Engineers in Newhall Street, to illustrate a paper read by the late Mr. Buckle, on Murdock's life and works. These experiments were not a source of unmixed gratification at headquarters, and we find Watt writing to Boulton: "I wish William could be brought to do as we do, to mind the business in hand." In 1785 he constructed a model of a steam engine with an oscillating cylinder. The invention of this type of engine has often been wrongfully ascribed to Richard Witty, whose patent only dates from 1813. Drawings of Murdock's locomotive and oscillating engine may be found in Mr. Muirhead's "Inventions of James Watt."

There is no reason to think that the interests of his employers suffered at Murdock's hands in consequence of these and other experiments, which must have been a relief from the constant worries and anxieties of his position. During his stay in Cornwall his attention was directed to the utilisation of iron pyrites, and in 1791 he obtained a patent for making sulphate of iron and a composition suitable for being used as a paint for protecting ships' bottoms and for other purposes. He proposed to roast arsenical pyrites either alone or mixed with zinc ore, and by properly adjusting the supply of air a sublimate will be obtained consisting of a mixture of orpiment, oxide of zinc, and sulphur. The sulphate of iron in the roasted mineral is dissolved out by water, and the solution evaporated to obtains crystals of the sulphate. The patent could not have been a valuable one at any time, and this is perhaps the reason why it has never been mentioned by any of Murdock's biographers. Curiously enough, it was referred to a month ago in a patent case in which the novelty of a certain arsenical paint was in dispute.

His investigations on the subject of coal gas, which were commenced at Redruth in 1792, were of far greater value, and in spite of the claims of the French, Murdock is entitled to the sole credit of being the first to bring gas-lighting into the region of practice. Part of the Soho establishment was regularly lighted with gas in 1798, and on the occasion of the rejoicings at the Peace of Amiens, in 1802, the front of the building was illuminated by the same means. In the following year the whole of the manufactory was lighted by gas. After that time it spread rapidly to other parts of the country. In 1808 he read a paper before the Royal Society, giving an account of his investigations on the subject, for which he received the Rumford Gold Medal. Murdock's researches appear to have founded a new branch of business at Soho—the manufacture of gas apparatus. We have never seen this fact recorded in any history of the establishment. Our authority for it is an incidental reference in a tract on gas-lighting, published by J. Maiben at Perth, in 1813.

We turn now from gas-lighting to Murdock's mechanical inventions. They are mostly comprised in his patent of 1799, which includes a rotatory engine, resembling a pair of accurately-fitting toothed wheels working together in a steam-tight casing. If high-pressure steam be allowed to act upon these wheels at their point of contact it is obvious that a rotatory motion will be the result. Murdock constructed a small engine upon this principle for driving the machinery in his private workshop at Soho. Another part of the patent is an improvement in steam jackets for cylinders, which had hitherto been built up in segments. Murdock's idea was to cast them in a single piece, in other words, to put the real cylinder inside another cylinder a little larger, so as to leave an annular space all round for the steam to circulate to keep the working cylinder hot. Another plan which he suggests is to cast the walls of the cylinder of additional thickness, and to bore out an annular space by means of a particular tool. This, it must be confessed, is not a very practical idea; but we must remember that in those days the difficulties of obtaining cylinders by casting were very great. (We shall shortly see how he made a further application of this principle some years afterwards.) His boring mill was a far more successful contrivance. Noticing the "jumping" which took place when a large cylinder was being operated upon, he proposed to drive the machine by a worm-wheel instead of a pinion, and the same mechanism was also applied to lathes for turning large works. Murdock was also the inventor of the slide-valve, the germ of which may be found in the patent above alluded to. It appears there in the form of a sliding eduction pipe, but we are unable to say when it developed into the long D valve, as it is technically called, from the resemblance of its cross section to that letter. The ordinary short valve, we may remark, was introduced about ten or fifteen years afterwards by Fenton, Murray, and Co., of Leeds. Murdock's valve was a great improvement upon those previously in use, and it was soon adopted by the Soho firm for their double-acting engines. He also devised a valve and gear not altogether unlike the Corliss valves of the present day.

In our opening article allusion was made to the extent to which Manchester had been indebted to other towns for the inventions on which its prosperity has been founded. In the year 1808 a body of persons, calling themselves the "Stone-pipe Company," obtained an Act of Parliament for supplying Manchester with water, and as their name imports, they proposed to convey the supply through stone pipes. One of the promoters, a certain Sir Geo. Wright, took out a patent in 1805 for a machine for making these pipes, the art of casting iron being, as we have before observed, in a very backward state. We do not know how Murdock's attention came to be directed to the matter, but in 1810 he obtained a patent for "boring and forming pipes, cylinders and columns, and circular discs, out of solid blocks and slabs of stone." In principle the machine consisted of a steel tube of the diameter of the inside of the intended pipe, having one end cut with teeth, forming what is known as a "crown saw." This tool, being presented to the end of a block of stone, and made to revolve, cut a hole clean through by removing a solid core, which in its turn was operated upon by a smaller saw, or was made to serve as a column for building purposes. There was a very ingenious arrangement for removing the chips by a current of water, which was caused to descend on one side of the tool and ascend on the other. The machine was tried at Soho, and subsequently at Rennie's works in London. It was eventually sold to the Stone-pipe Company, and used for making the tubes laid down in Manchester. We may here remark that one William Docker, who carried on the business of a slate merchant in Birmingham, had a patent in 1798 for stone pipes, the principal feature of which was a particular form of joint. Perhaps we ought to add that stone water pipes turned out a decided failure.

Murdock was strongly impressed with the value of compressed air as a motive power, and he applied it for driving the lathes in the pattern shop, and for actuating a lift at Soho; and his house at Sycamore Hill was fitted up with a very ingenious system of pneumatic bells—a contrivance afterwards imitated by Sir Walter Scott at Abbotsford. The system of transmitting parcels through an exhausted tube, which has since his day come into very general use under the name of the Pneumatic Despatch, was also devised by him. The mixture of iron borings and sal ammoniac, known to every engineer as "iron cement," was also due to him, although it appears that Watt made the discovery about the same time. Murdock was also the inventor of that elegant piece of mechanism known as the "sun and planet motion," a substitute for the crank which the firm were compelled to use in consequence of Pickard's patent for the application of the crank to the steam engine. The contrivance is included in Watt's patent of 1781, but it was really the product of Murdock's ingenuity, as Watt himself states. The idea of using fish skins as a source of gelatine to be employed in clarifying beer has been attributed to Murdock, but he was not the first to suggest the application, as it formed the subject of a patent as early as 1760.

Murdock died at his house at Sycamore Hill, Handsworth, on the 15th November, 1839, in his 85th year, and is buried in Handsworth Church, not far from Boulton and Watt, whom he faithfully served for more than fifty years.

We bring our sketches of the Soho group to a conclusion with a brief notice of

## John Southern, F.R.S.

**Southern was not, strictly speaking, an inventor, but we have included him in this series from a feeling that his**

talents have not been done justice to. He was born in 1758, in the neighbourhood of Birmingham, we believe. When he was twenty-two he entered the service of the Soho firm, in whose employ he continued until his death in 1815. In his earlier years he seems to have been looked upon as rather "flighty," and we find an uncomplimentary reference to his musical tastes in a letter from Watt, dated October 1, 1781, quoted by Mr. Muirhead, to the effect that Southern must "give bond to give up music; otherwise I am sure he will do no good, it being the source of idleness. He was an excellent mathematician, and he was very useful to Watt in that capacity. His researches on the elasticity, density, and latent heat of steam, which were undertaken at Watt's request in 1803, were for a long time the standard authority upon the subject. They are printed in Brewster's edition of Robison's "Mechanical Philosophy," accompanied by a series of formulæ deduced from the experiments. He was the author of a tract on balloons, published at Birmingham in 1785, at the time when the ascents by the French aeronauts were exciting much attention. The book contains practical directions for making balloons, as well as some theoretical investigations on their ascensional power. He also contributed to the "Philosophical Magazine" papers on the equilibrium of arches (1802), and the results of some experiments on friction (1803). He died at Oakhill, Handsworth, July 28, 1815, aged 57, and is buried in King's Norton Church, in the family vault of his father-in-law, Thomas Dobbs, whom we shall have occasion to speak of as an inventor in a future article. The inscription on his monument says :—"He was distinguished by his extensive and accurate acquaintance with the mathematics and the different branches of physical science, and by the singular ability with which he applied his knowledge to important practical purposes during a connection of thirty-five years' continuance (and terminated only by his death) with the celebrated establishment of Soho. The integrity of his character and his affectionate and faithful discharge of the social duties of life secured to him the attachment of many friends, and caused his name to be cherished with the fondest recollection by his afflicted family."

## CHAPTER VI.

JAMES WATT'S INVENTIONS—SEPARATE CONDENSER—HIGH PRESSURE AND ROTARY ENGINE—COPYING MACHINE—CRANK INVENTIONS—MURDOCK'S SUN AND PLANET WHEEL.

### JAMES WATT.

Without being open to the charge of making sweeping accusations, we venture to say that one of the prevailing characteristics of the present age is ingratitude to the memory of its greatest benefactors. True, when our attention is directed to the matter, we acknowledge in a careless, general way that certain names are connected with certain inventions, but there the matter ends, and this is particularly the case with everyday contrivances. Familiarity has in truth bred a certain amount of contempt, or rather, perhaps, indifference. We enjoy the benefits, caring little to enquire how they came to us, and whose was the brain that contrived them. In the words of Lord Brougham's noble epitaph on the monument in Westminster Abbey, the name of Watt "must endure while the peaceful arts flourish," and it is indissolubly connected with the steam engine. The labours of James Watt have no need of any further eulogy, and it would be impertinent to attempt to do over again what has already been done so well by more than one writer. But a necessity does exist for a brief chronicle of the details of his great work, and every Birmingham man who takes an intelligent interest in the history of the town, when asked "Who was James Watt?" should be able to say more than that he was "the inventor of the steam engine, and you'll find his monument just by the Midland Institute." We have seen from time to time some curious statements on this question. A pamphlet dealing with the gas question was issued the other day, and the author prefaces his remarks on the gas regulator in the following extraordinary way:—"The great Watt, to whom the engineering world is so greatly indebted, when a lad had to attend to a pumping engine, the speed of which he was instructed to control by opening or shutting the steam-valve, according to its requirements; this he did by knocking the steam cock with a wooden stick. But his ingenious and active mind soon found a substitute for that tedious and somewhat primitive occupation, and the steam governor, so extensively used now, was the result of his conception." This is a clumsy version of the story of Humphrey Potter, a boy who is said to have invented a rough kind of self-acting valve-gear for pumping engines. It is easily disposed of by the simple face that Watt was some 25 years of age before he saw even a model of a steam engine. Again, in a popular magazine for children, we find a picture of "Watt and the steam," in which a semi-idiotic youth is represented watching a boiling tea-kettle "in his father's poor little home," as the explanatory article says. We forgive the writer for his inaccuracies on account of the advice which he gives to his young friends to "use their minds," and "to think about things. Ask yourself how this is done and why. Ask other people, and when you have found out, don't forget all about it; but remember it, think over it, and use it."

It is time, however, to commence an account of the inventions really due to Watt; and first in point of time as well as importance we must rank his discovery of separate condensation, for which he obtained a patent in 1769, when he was a mathematical instrument maker in Glasgow. In the machines made up to that date condensation was effected in the cylinder itself, and a great loss of heat ensued by the quantity of steam which was wasted when a fresh stroke was about to be made. It will be obvious that it is impossible to fill a vessel full of steam unless the vessel is of the same temperature as the steam itself, otherwise condensation will take place. Watt conceived the happy idea, or inspiration shall we call it, of using a separate vessel in which to condense the steam and form the necessary vacuum. The patent also includes the well-known "steam jacket" for preventing loss of heat by radiation from the cylinder, the air-pump, the use of high-pressure steam, a particular form of rotary engine, and a mode of keeping the piston tight by the use of oil, fat, wax, resinous bodies, or mercury. Such very briefly is a sketch of a patent which laid the foundation of the steam engine of our day. After many weary months of experimenting, in which he was magnificently assisted by his friend Roebuck, who had then settled in Scotland, he succeeded in bringing the machine into a practical shape. In due course he was introduced to Boulton, and removed to Birmingham.

Pursuing the subject in chronological order, a period of eleven years elapses before we find Watt's name again in the patent records. The subject appears to be singularly wide of that with which his name is most generally associated, and of the thousands who avail themselves of the copying press, how many, or rather how few, know that they are indebted for it to the versatile genius of James

Watt. It was patented February 14, 1780, but the inventor had perfected it about two years before. The machine is too well known to require description, but it may be well to put on record the recipe for the ink most proper to be used. It is as follows :—"Take 4 quarts, ale measure, of spring water, 1½lb. Aleppo galls, ½lb. green copperas or green vitriol, ½lb. gum arabic, 4oz. roche alum ; pound the solid ingredients, and infuse them in the water six weeks or two months, during which time the liquor should be frequently shaken. Strain the liquor through a linen cloth, and keep it in bottles closely corked for use." Here is an idea for an enterprising ink maker—"The Original Copying Ink. Warranted prepared from James Watt's own recipe." Two kinds of presses are described, a rolling press somewhat similar to that used by copper-plate printers, and a screw press like those now universally employed. Watt preferred the former construction, and some of the original machines are in existence, and perhaps in use at the present day. We believe that the roller machines are still made at Soho, being preferred for certain kinds of work. Some difficulty was experienced at first in getting the process introduced, as it was feared that it would facilitate forgery. It must, however, have become rapidly known, for only five years afterwards we find it referred to incidentally in a patent granted to William Alston, of Birmingham. The patentee was desirous of indicating the nature of a certain machine which he used for rolling metals, and he compares it to "Mr. Watt's copying-press," so that we may conclude that it was tolerably well known by that time. There is one part of the specification which has not stood the test of practice—that in which the inventor describes a special liquor to be used instead of water for damping the paper, for the purpose of bringing out the writing more distinctly.

Watt's next patent was granted in 1781. It is only of passing interest, and would never have been taken out but for the fact that Watt conceived himself shut out from using the crank by Pickard's patent of August 23, 1780. The specification includes five different pieces of mechanism for converting the vibrating motion of the beam into a continuous rotary motion. The first is a modification of the "swash-plate," as it is technically called, the construction of which may be found in any book on mechanical motions. The second consists of an eccentric keyed on the main shaft of the engine, encircled by what is in fact the equivalent of the ring or "strap" now used. The third and fourth are nothing more than thinly-disguised cranks, and their use would certainly have rendered Watt liable for an action for infringement had he employed these contrivances. The fifth was the well-known "sun and planet" wheel, the invention of Murdock, as we showed last week. We believe we are correct in saying that this was the only part of the patent under notice which was ever brought into practical use, and it was abandoned on the expiration of Pickard's patent. A sun-and-planet engine might, however, have been seen at work at Soho about 25 years ago, and "Old Bess," as she was called, has found an honourable resting place at the Patent Office Museum, South Kensington. The Department of Science and Art is the fortunate possessor of some interesting contemporary models of the various mechanisms described in this patent. They were presented to the Department by Mr. Gilbert Hamilton, after having been exhibited at the Loan Collection of Scientific Instruments in 1876.

Allusion has been made above to a prior patent for the crank by James Pickard, a Birmingham button maker, and Mr. Smiles does not hesitate to describe him as a thief who deliberately stole the invention. In the "Lives of Boulton and Watt" there is a striking picture of the interior of the "Wagon and Horses" at Handsworth, an inn where the Soho workmen were wont to meet and discuss the doings of the firm. Pickard appears as a mysterious stranger, listening to the conversation, and the story goes that "he posted straight to London and anticipated Watt by securing a protection for the contrivance." There are very grave doubts as to the truth of all this, and it is highly probable that Pickard was an original inventor. It is absolutely certain that he was at work upon the subject early in 1779, and he actually applied for a patent for a machine of the kind for Scotland in that year. This fact is now published for the first time, but unfortunately no details of this invention are extant, as the patent was abandoned, and consequently no specification was enrolled. He was in communication with Matthew Wasbrough, of Bristol, a very ingenious mechanic, who died too early for fame, and who was also the patentee of mechanism for obtaining rotary motion in steam engines. Mr. Matthew Wasbrough, grandson of the engineer, the present coroner for the city of Bristol, has in his possession a deed, dated April, 1779, from which it appears that Pickard, Wasbrough, Sampson Freeth—a Birmingham merchant whom we shall have occasion to speak of in a future article—and William Chapman, of Newcastle-on-Tyne—a celebrated engineer of the last century

—were in partnership for the purpose of introducing a certain contrivance to be applied to steam engines for obtaining rotary motion. Is it not probable that Pickard's alleged fraudulent patent of 1780 may have been the natural outcome of his experiments? There is, it must be admitted, some evidence of his want of *bona fides* in the variations which exist in the title of his invention at the different stages through which a patent had to pass in those days. In his initial application it is described as "a new-invented mill or machine, upon an entire new construction, which will be of great advantage in various manufactories, such as rolling, turning, boring, grinding of corn, and every other work that a mill is capable of performing by a rotative motion." This looks perfectly innocent, and would not have aroused the suspicion of Watt, who was always on the alert; but when the time arrived for filing the specification the mask was thrown off, and the title was altered, so as to fully indicate the nature of the invention. It is stated in C. Pye's "Birmingham in 1818" that the first crank engine ever erected was that put up by Pickard, in a corn mill at the bottom of Snow Hill, in 1781. The author continues, "I do not know that he ever erected any others; for Messrs. Boulton and Watt, in order to evade the patent (!), substituted the sun-and-planet wheels, which they continued to use until the patent expired." By way of conclusion to this episode, we may remark that it seems to have escaped notice that the application of the crank to the steam engine is included in the specification of a patent granted in 1766 to John Barber, of "Standsby, in the county of Derby"—probably an error for "Stainsby Hall," in the same county, as given in Barber's later patents. It is also mentioned in a patent for working coal mines, granted in 1761, to Michael Meinzies, the Scotch advocate who gave us the thrashing machine. The crank is there suggested in the clearest manner for working a coal-cutting machine. "The fire engine," he says, "and the other machines can communicate their power by the up-coming and down-going of spears in the coal-pit, not only by moving an axletree, with a double crank, in the manner above set forth, placed either at the surface or near the bottom of the pit," but by other mechanism which he proceeds to describe.

We shall continue the account of James Watt's inventions next week.

## CHAPTER VII.

James Watt's Inventions—Use of Steam Expansively—The Parallel Motion—Steam Hammer Valve—Projected High-pressure Locomotive—Smoke-consuming Furnace—Centrifugal Governor — Water Gauge — Engine Counter — Sculpture-copying Machine — Letter-copying Machine—Flexible Water-main—Discovery of the Composition of Water—Remedial Gases in Disease.

### James Watt—(concluded).

We have now arrived at that which we venture to regard as Watt's second great patent, that of March 12, 1782, which includes the use of steam expansively—an idea which had long been present to Watt's mind. The use of high-pressure steam, cut off before the piston has arrived at the end of its course, was a great stride, and has been largely developed since the period of its introduction. The double-acting engine, which also forms part of this patent, "consists in employing the elastic power of the steam to force the piston upwards and also to press it downwards alternately by making a vacuum above or below the piston respectively, and at the same time employing the steam to act upon the piston in that end or portion of the cylinder which is not exhausted, so that an engine constructed in this manner can perform twice the quantity of work or exert double the power in the same time (with a cylinder of the same size) which has hitherto been done by any steam engine in which the active force of the steam is exerted upon the piston only in one direction, whether upwards or downwards." Another feature of this patent is the compound or combined high and low pressure engine, in which the waste steam is passed into a second cylinder, and from thence into the condenser. This principle has been greatly extended, and has led to a most important economy in the consumption of fuel, though it is only within the last few years that the importance of its application to marine engines has been completely recognised. We have also a method of connecting the beam with the piston-rod, consisting of an arrangement of racks and sectors, but which the invention of the parallel motion soon rendered obsolete. The specification concludes with a description of a rotary engine, which probably succeeded no better and no worse than the hundreds of plans which have been proposed since that period.

Few persons, we think, will be inclined to agree with Mr. Muirhead in his estimate of the importance of the patent of 1784, which he ranks immediately after the separate condenser patent of 1769. Beyond that exquisite piece of mechanism, the "parallel motion," it contains little of permanent value. The first portion of the specification treats of a particular form of rotary engine, on the principle of the well-known æolipile. The second part includes three modes of connecting the extremity of the working beam, which, of course, moves in the arc of a circle, to the extremity of the piston-rod, the motion of which is rectilinear. Amongst these we find the "parallel motion," of which the inventor said, "Though I am not over anxious after fame, yet I am more proud of the parallel motion than of any other mechanical invention I have ever made." The third part embraces various methods of connecting the pump-rods to the beam, and the fourth improvement is a mode of applying the power of steam-engines to work rolling mills, &c. In the fifth part we find the *direct-acting* steam hammer of the present day accurately foreshadowed, but not described in detail. Watt proposed to fix "the hammer or stamper either directly to the piston or piston rod of the engine, or upon or to the working beam of the engine, or by fixing the hammer or stamper upon a secondary lever or helve, and connecting the said lever or helve, by means of a strap or strong rod, to or with the working beam." The latter is the form described in the specification. There is also a quick valve motion for use in a steam hammer, or whenever a very rapid action is required. The drawings show a "trunk" or hollow piston-rod, but there is no allusion to it in the text; and curiously enough no claim is made to the contrivance, but we shall hardly be wrong in attributing the invention to Watt. The trunk engine was patented by Humphrys in 1835 as an entirely new thing. The specification concludes with a description of a locomotive, not, however, of a very practicable character, for it includes apparatus for condensing the steam. This was a branch of the subject which Watt does not seem to have pursued any further, and we believe that he never constructed a model of his invention.

The last patent taken out by Watt is dated June 14, 1785, and is for improvements in furnaces for steam boilers, &c., and for metallurgical purposes. The principle consists in "causing the smoke and flame of the fresh fuel to pass through very hot tunnels or pipes, or among, through, or near fuel which is intensely hot, and which has ceased to smoke, and by mixing it with fresh air when in these circumstances." This is the foundation of nearly all the contrivances which have been put forward for "consuming smoke," an expression which occurs for the first time, we believe, in the specification of the patent in question.

Amongst Watt's unpatented inventions we may mention the governor and throttle valve, for automatically regulating the supply of steam to the engine. Attempts have been made to deprive Watt of the credit of this beautiful contrivance, by citing a patent granted in 1787 to Thomas Mead, where the centrifugal principle is applied to a mechanism for furling and unfurling the sails of windmills, thus exposing a smaller or larger surface to the wind, according to the quantity of corn passing between the stones. But Watt is certainly entitled to the credit of a new application of a well-known principle. Engineers will appreciate the value of such contrivances as the stuffing box, the water gauge for boilers, and the cone at the end of the piston-rod for fastening it to the piston. But, perhaps, the most exquisite piece of mechanism of all is the indicator, which has been well compared to the stethoscope of the doctor. It enables us in truth to know what is going on within the cylinder--whether the valves are working properly, and whether any leakage detrimental to the action of the engine is taking place. Watt kept the invention a secret for many years, and in 1814 when he published an account of his miscellaneous inventions in Brewster's edition of Robison's "Mechanical Philosophy," he only gave an imperfect description of this, without any hint of certain essential parts. A complete instrument turned up in Russia, where it had been made by some of the people sent out by Boulton and Watt to erect a steam engine in that country. This fell into the hands of the late John Farey, a well-known civil engineer, the author of a work on the steam engine. On his return to this country, as he stated in his evidence before a Parliamentary Committee on the Patent Laws, he made an indicator after the pattern of that which he saw in Russia, and "showed several other engineers how to make such for themselves, and since that time every one of these persons has very greatly improved his practice by the light that it has enabled him to throw upon an obscure part of the operation of steam in a steam engine." Watt was also the inventor of the "engine counter," a machine which, being attached to some moving part, shows the number of strokes made by the engine. Watt's letters to his friend Small, published in Mr. Muirhead's book, abound with pen sketches of improvements relating to numberless subjects. In his early days he invented a machine for drawing in perspective, and during his retirement at Heathfield he was diligently at work upon a sculpture copying machine, the remains of which are still in the "classic garret" at that place. When he was seventy-five years of age he designed a water main for conveying water across the Clyde, fashioned after the manner of the tail of the lobster, and possessed of sufficient flexibility as to allow it to adapt itself to the shifting river bottom.

Watt's researches in pure and applied chemistry yet remain to be alluded to. In the words of Arago, the discovery of the composition of water is "one of those discoveries which do the greatest honour to the human intellect." The honour has been hotly contested, but the distinguished French philosopher and Lord Brougham both unite to award it unhesitatingly to James Watt, who first communicated his theory in a letter to his friend Dr. Priestley, on the 26th of April, 1783. Then, he was one of the first, if not the very first, to introduce the art of bleaching by chlorine into this country, which he had learned from the inventor, Berthollet, a distinguished French chemist, during a visit to Paris in 1786-7, and shortly after his return home we find him in communication with his father-in-law, Mr. Macgregor, on the subject of the practical application of the new agent. He also devoted much attention to the subject of the remedial action of gases in various forms of disease. The idea was started by Dr. Beddoes, of Bristol, and Watt entered into it with his usual enthusiasm. He devised several forms of apparatus for making and respiring the various gases, and embodied the results in a pamphlet. The apparatus was made at Soho, and no doubt eventually led to the manufacture of appliances for producing coal gas, as noticed in our article on Murdock.

Such is a very imperfect, and we fear rather dry, summary of Watt's inventions. No attempt has been made to trace the steps by which they were gradually brought to perfection, nor the difficulties which he met with in introducing them into actual practice. To do justice to the ingenuity which they display would require a volume, rather than a couple of articles; but it was absolutely necessary to include some account of the work of the most philosophical inventor that the world ever saw in this series. Watt's inventions have exercised so vast an influence upon the world that it has been customary to regard them as belonging rather to mankind in general than as associated with any particular place. But when we think of the material assistance and never-failing encouragement which Watt received from his courageous and sanguine friend, we may well say that, if there had been no Boulton, there would have been no Watt. Without unduly pressing what has been called the "parochial idea," it may be affirmed that if there had been no Birmingham there would have been no Boulton.

## CHAPTER VIII.

BASKERVILLE AND HENRY CLAY—CLAY'S ORIGINAL PATENT—PAPER FURNITURE—PAPER HOUSES, SHIPS, BOATS, &c.—THOMAS JONES AND BRINDLEY'S PATENTS —VARIOUS PROCESSES—JENNENS'S PATENT.

### PAPIER MACHE.

Those who are in the habit of regarding words as something more than mere arbitrary signs, like the numbers in a telegraphic code, must be frequently puzzled to account for the apparently conventional character of much of our technical nomenclature. For instance, "black lead" appears to have been so designated on the *lucus a non lucendo* principle, seeing that lead does not enter into its composition at all; "Berlin wool" is totally unknown at Berlin; and the words "papier mâché" suggest to the ear of a Frenchman anything but a substance suitable to be employed where considerable strength is required. In fact, the great French dictionaries, such as that of Littré, do not give the word at all, except in a figurative sense expressive of the last degree of weakness, such as that ascribed to wet blotting-paper. Nor is it to be found in many of their technical dictionaries, and in Laboulaye's "Dictionnaire des Arts et Manufactures," a well-known work, which has passed through several editions, it appears for the first time in the supplement published in 1875. The author says: "The processes of this manufacture are little known, and we met with a description for the first time in a work by M. Jobard." The work referred to is "Nouvelles Inventions aux Expositions Universelles, 1858," where we read "that the name papier mâché, given by Messrs. Jennens and Bettridge to their charming productions, is incorrect. It should have been called papier collé," that is, "pasted paper," or pasteboard. This is a perfectly just criticism, as regards articles made by attaching sheets of paper together, but the writer had forgotten that the earliest papier mâché consisted of paper pulp, cast into shape in moulds. This process has been known for two centuries, and it is mentioned by Robert Boyle in his essay "Of Man's great Ignorance of the Uses of Natural Things" (1672), where he says: "Though paper be one of the commonest bodies that we use, there are very few that imagine it is fit to be employed other ways than in writing, or printing, or wrapping up of other things, or about some such obvious piece of service, without dreaming that frames of pictures, and divers fine pieces of embossed work, with other curious movables, may, as trial has informed us, be made of it."

The earliest detailed description of the pulp process we have met with is that contained in the "Complete Dictionary of Arts and Sciences," by the Rev. T. H. Croker and others, published in 1765. It was probably introduced into Birmingham about the middle of the century by Baskerville, whose fame as a maker of japanned goods we have already referred to. In 1770 the manufacture was so thoroughly established as to have become a separate trade, and we learn from Sketchley and Adams' "Directory" for that year that Watson and Co. carried on the business of "papier-machie manufacturers" at No. 76, High Street.

Amongst Baskerville's apprentices was a youth named Henry Clay, who, when he had served his time, set up in business on his own account in partnership with Mr. Gibbons, in Newhall Street, according to the Directory quoted above. Clay conceived the happy idea of using sheets of paper pasted together instead of pulp, and he took out a patent, for "making in paper high varnished panels or roofs for coaches, and all sorts of wheel carriages and sedan chairs, panels for rooms, doors, and cabins of ships, cabinets, bookcases, screens, chimney pieces, tables, tea trays, and waiters," on the 20th of November, 1772. As this patent was the foundation not only of the fortunes of the inventor but also of an important Birmingham trade, we give the specification in full:—

"My inventions of making paper panels for various purposes are by pasting several papers upon boards or plates of regular thicknesses on each side the same, to prevent one side counteracting or drawing with superior force to the other in the state of drying; and when the same is rendered sufficiently strong for the purpose intended it is then planed or cut off at the edges until the board or plate appears, and then taken off such boards or plates. The pieces of paper are afterwards screwed or fastened on boards or plates, and put in a stove sufficiently hot to deprive them of their flexibility, and at the same time are rubbed over or dipped in oil or varnish, which so immediately drenches into them as to secure them from damp, &c. After the papers are thus made they are capable of being sawed into different forms, and planed as wood; and if joining the parts be required, as in tea trays, dressing-boxes, and some other articles, they are done by dovetailing or mitreing, in the same manner as if made in wood. After the various articles are thus formed in the paper, they are then coated with colour and oils sufficient to make the surface even, and are then japanned and high varnished."

The process is simplicity itself, but excellence is only to be attained by careful attention to details, such as the quality of the paper, composition of the varnish, temperature at which the articles are dried, and so forth. Clay understood this well, and his goods were triumphs of the japanner's art. Some of them are said to be in existence and in use at the present day. He was fortunate in obtaining the patronage of Royalty, and we read in the local paper how, on the 27th of May, 1793, "Mr. Clay, of this town, had the honour of introducing to their Majesties and the Princesses, at the Queen's Palace, a new material for painting upon infinitely superior to any substance hitherto used for its surface and durability." The Queen subsequently accepted a present of a sedan chair, the panels of which were of papier mâché, and a pair of console tables adorned with paintings after Guido. Clay made a large fortune, and in 1790 he was chosen to fill the office of high sheriff of the county. On the expiration of the patent several houses started the manufacture, but the original business appears to have survived somehow, though we are unable to trace its descent, and papier mâché articles were shown at the Great Exhibition of 1851 by W. Clay and Son, of Fenchurch Street, London, who claimed to be the representatives and descendants of the first inventor.

Notwithstanding the widespread publicity which Clay's invention obtained it is somewhat curious that no account of the process seems to have found its way into books for nearly half a century afterwards. We have searched most of the great encyclopædias published in the early part of the present century, and although the word occurs in several of them, it is the old pulp process which is described, and not that patented by Clay. It will be observed that he does not use the word "papier-mâché" to denote the new material, nor does it occur in any of the patents having reference to the subject for many years afterwards. May it not have been that the word is of comparatively recent introduction as applied to goods manufactured by pasting sheets of paper together? This suggestion is partly borne out by the fact that in Bisset's "Magnificent Directory" Clay simply appears as "Japanner in ordinary to his Majesty." It is true that the "Japan and Paper Machee Manufactories" figure among the sights of the town in the same work, but the old process may be meant.

A rather remarkable patent was taken out in 1788 by one Charles Lewis Ducrest, of Jermyn Street, London, which calls for some notice here, although not relating to the town. The title is "Making paper for the building of houses, bridges, ships, boats, and all sorts of wheel carriages, sedan chairs, tables, and book cases, either of paper or wood, and iron covered with paper." Carriages and sedan chairs were made by pasting sheets of paper over a wooden mould, which was afterwards taken to pieces and removed. Minute directions are given for attaching the axletrees, hanging the doors, and finishing, varnishing, and stoving the body. Furniture is made of light white wood, covered over with successive sheets of paper, and boats are made entirely of that material. There is rather an original way of constructing houses, each room being made separately, like a box, and fastened to its neighbour by iron bolts. The inventor does not confine himself to a single storey, but piles these boxes one above another, suitable ties being provided to give rigidity to the structure. Bridges are made of ropes covered with paper, the roadway to be of wood or iron plates, coated with paper. The detail is worked out with such minuteness that it has every appearance of a serious and practical proposal. Clay's patent had expired, otherwise it would have been regarded in the light of an infringement.

With the exception of three patents for making buttons of papier-mâché, which will be described further on, no patent appears to have been issued until the year 1805, when protection was granted to Thomas Jones, japanner, of Bilston, for a method of making trays, &c., from sheets of wet pulp, by pressure between shaped moulds. The process cannot be distinguished from one patented forty years afterwards, now to be described. In 1836, Mr. William Brindley, a papermaker, carrying on business in Caroline Street, took out a patent for making trays, &c., by piling sheets of paper wet from the pulp vat until a sufficient thickness had been obtained. The superfluous moisture was squeezed out by pressure between felts and the pile was placed in a shaped metal mould, where it was subjected to further pressure, and dried without being removed. The process was found to be objectionable on account of the fractures which took place in the compound sheets, and the inventor took out a subsequent patent in 1845, according to which he employs a solid sheet of pulp like that prepared when making millboard. This is afterwards pressed and stoved between dies. It would require considerable ingenuity to point out the difference between this process and that invented by Jones in 1805. Mr. Aitken states that Brindley's first patent was declared void, but we are unable to refer to any report of the proceedings. It was a partial return to the old process of casting fibrous pulp in moulds alluded to by Boyle : but we believe that Brindley was the first to show practically how large flat articles, such as tea-trays, could be made in this manner. Obadiah Westwood, a button-maker in the town, had indeed proposed to make them in this way in his patent of 1786, but he merely mentions them in the title of his patent, without giving any description.

We may here allude to Mr. Theodore Hyla Jennens's patent of 1847 for softening blanks by steaming before they are pressed, and for an improvement in the moulds employed. Up to that time moulds had been made upon an erroneous principle, the top and bottom moulds having been constructed to fit each other. This was obviously wrong, as will be evident from a glance at the section of, say, a card-basket. If the article has an equal thickness throughout, which it was Mr. Jennens's object to produce, the contour of the inside is not identical with that of the outside. The patentee shows the tearing action which takes place when using the old-fashioned moulds by means of diagrams, but the wonder is that it was not observed long before, especially in a town where the operation of moulding for metal casting is so largely practised. Simple as the invention is, it was nevertheless repatented in 1855 by F. D. Blyth.

Next week we shall deal with the various methods of ornamenting papier mâché, and describe some miscellaneous applications of the material.

## CHAPTER IX.

### PAPIER MACHE—(CONTINUED).

DECORATION PROCESSES — TRANSFERRING — PEARL-INLAYING — GEM-INLAYING — ELECTROTYPE ORNAMENTS — USE OF ORNAMENTAL MOULDS — MISCELLANEOUS APPLICATIONS OF THE MATERIAL.

With regard to the ornamentation of papier mâché goods it is obvious that the processes used for decorating japanned articles are in general applicable also to the first-named material. At first painting by hand and gilding were employed, and as the demand for decorated articles increased mechanical means of multiplying the designs were sought for, and the transfer process was invented. This was no doubt suggested by the method employed for decorating earthenware, first practised secretly about 1756 or 1757. A very interesting attempt in this direction was made by one Stephen Bedford, a Birmingham japanner, in 1759, when he obtained a patent for "impressing in imitation of engraving upon varnish laid upon copper, iron, *paper*, and other bodies, to be used in coach panels, snuff boxes, and other kinds of merchandise, impressions of foliages, figures, decorations, ornaments, and other devices." The plan was, it must be confessed, not very practical, for it consisted in transferring impressions from copper plates printed on thin sheet lead. The lead would certainly stretch and thus produce a blurred impression, and the metal would not yield up the colour very readily, but still we have the principle of transferring clearly indicated. We take the word "paper" in the title to mean articles of solid paper pulp, as there could be no reason for adopting a roundabout method of ornamenting single sheets, when the result could be more easily achieved by printing direct from the plate. The patentee appears to have carried on business for many years afterwards, and he figures amongst the japanners in the "Directory" for 1770, his address being No. 135, Moor Street. There was a "Stephen Bedford, ironfounder"—perhaps a son—living in Birchall Street in 1819, in which year he took out a patent for making iron.

The transfer method was still further developed, and indeed perfected, as far as theory goes—for success depends very largely upon manipulative skill—by one Charles Valentine, japanner, of Clerkenwell, in 1809. The design is to be engraved on as many plates as there are colours in the picture—he mentions four in his specification—and an impression is taken on thin paper prepared with isinglass or any glutinous substance to prevent the ink from sinking in. This is then to be transferred to the goods, and the paper removed by damping, leaving "the finished painting of four shades." He obtains designs in gold by first gilding the entire surface of the article, and securing it by a coat of varnish, which must be allowed to dry. Then transfer an impression of the required design printed with a mixture of printer's ink and wax. The ground is next to be covered in with opaque varnish—the pattern of course being left intact—and when this is dry the printer's ink may be removed by warmth and friction, revealing the pattern in gold underneath.

In 1844 Mr. Thos. Farmer, of this town, patented a slight modification of the above process, the only difference being that the gold leaf not required to form the pattern is removed by friction instead of being covered up, and the varnishing takes place at a later stage. Mr. Charles Breese, of the firm of Breese and Hayward, repatented the same process in 1853, and Mr. George Haseler, of Wolverhampton, appears in the following year as the patentee of a method by which the parts *not* required to be ornamented were printed with varnish, the plain portions being gilt, silvered, or bronzed by any of the known methods. The varnish is subsequently removed by a solvent.

Mr. Breese also patented in 1853 a process for ornamenting by transferring a pattern printed in adhesive varnish and dusting the metal on in a powdered condition, or applying it in leaf.

Mr. George Goodman, jun., in 1852, and Messrs. Samuel Tearne and George William Richmond, in 1856, appear as patentees of a method of transferring pictures in several colours which is not distinguishable from that described by Valentine in 1809, except that lithography is employed as a means of obtaining the pictures to be operated upon. The Mr. Richmond above mentioned is, or was, a lithographic artist in the employment of Mr. Tho: Underwood.

With regard to what may be called the *appliqué* style of ornamentation, the material most commonly used is pearl shell, which was mentioned, as we shall see, by Clay in his patent of 1778, and again by Westwood in 1786, but it was not until 1825 that the decoration of papier mâché formed the subject of a separate patent. In that year Aaron Jennens and John Bettridge, who are described as "Japanners in Ordinary to his Majesty," obtained protection for a very ingenious process for obtaining devices in pearl shell suitable for inlaying. We use the word "inlaying" for want of a better, although it is not strictly correct, the pattern being only stuck on to the surface of the material. The pearl shell having been ground sufficiently thin, the device, be it flower, leaf, star, or what not, is traced out in etching ground or varnish. The pearl is then lightly brushed over with strong acid, which soon eats its way through, and thus separates the device protected from the action of the acid by the varnish. Sometimes the whole of the plate is covered with a "ground," which is removed by the ordinary etching needle in those places where it is desired to separate the material. The patentees also state that a penknife or proper cutting tool may be used, but the etching process produced by far the most delicate work. When a large number of precisely similar patterns are required, layers of shell are united with cement. The composite block is then filed to the shape of the required ornament, and when it is finished the cement is dissolved, the layers separate, and we have a number of devices all exactly alike. Mr. Aitken, in his article on the papier mâché trade in the "Industrial History of Birmingham," states that this was invented by George Souter, a decorator in the employ of Jennens and Bettridge.

It was proposed nearly a century ago by one John Skidmore, of Coppice Row. Clerkenwell—who, we cannot help thinking, must have been a progenitor of Skidmore of Coventry—to ornament "all sorts of japan wares, all kinds of cabinets and furniture, &c., &c.," with "Bristol stones, paste, and all sort of pinched glass, sapped glass, and every other stone, glass, and composition used in or applicable to the jewellery trade." Mr. Theodore Hyla Jennens patented the application of the same materials in 1847. Mr. Aitken states that this process was the invention of Benjamin Giles.

A highly ingenious method of ornamenting was included in Mr. Farmer's patent, already referred to. He proposed to deposit copper by the electrotype process upon a pattern traced out in gold or silver leaf, or it is presumed in any medium possessing the property of conducting electricity. When the deposit has attained the thickness of about 1-60th of an inch, it should be varnished and then ground down until the copper is again exposed. More copper is deposited and then gilt or silvered according to taste. Detached ornaments may be made by the electrotype process and applied in the same manner as those of pearl shell.

Ornaments in relief are produced by simple pressure in moulds in which the required design is sunk, according to the method patented by Mr. Brindley in 1848, the material being in a wet state. To save the wear of metal moulds, embossed sheets of papier mâché, stoved and varnished, are found sufficiently durable to give the requisite impression. The patent granted to Mr. Brindley in the following year contained a method of using leaves, flowers, and other natural objects for obtaining ornamental effects on papier mâché. This is an application of "nature-printing," which we have never seen alluded to in any of the many discussions upon the subject.* A more elaborate method of ornamentation was that included in Mr. Farmer's specification, according to which the design is first moulded in plaster of Paris or wax and an electrotype taken. The mould is then covered with several coats of japan, and layers of pulp are added until a sufficient thickness has been obtained. The copper mould is removed by means of acid, and a fine sharp reproduction of the original is the result, the thick coating of japan having received the impression of the mould. The process must however have been expensive.

We must not omit to mention a curious patent granted to Samuel Hands, a Birmingham buckle maker, in 1790, for ornamenting articles of leather or paper (that is solid paper). The specification is rather vague, but it is so far clear as to be evident that he proposes to ornament papier mâché by embossing with suitable engraved dies, and also by inlaying with various ornamental substances, the ornaments being attached by cement, riveting, or by simple pressure. The patentee appears to have been an importation, if we are right in assuming him to be the same Samuel Hands, of Wirksworth, who had a patent for buckles a few months previously. The process may not have been of much practical importance, but it serves to indicate the direction in which men's minds were working.

As we showed last week, it was proposed by Ducrest as early as 1788 to make furniture by coating a wooden body with successive layers of paper, but we are unable to fix the date when the manufacture of such articles was commenced. A papier mâché ottoman was made for the Duchess of Sutherland in 1840, by a Birmingham firm, and was described in a paragraph in the "Mechanics' Magazine," for July 11 (quoted from the *Staffordshire Examiner*), as being "equally beautiful and novel." This date may perhaps mark the commencement of a trade which might have had a permanent success had the principles of decoration and constructive design been as well understood as were the processes of making the raw material. Joseph Bettridge, wood turner, patented in 1841 a

* The late Mr. Richard Ford Sturges patented the application of nature printing to the ornamentation of metals in 1852.

method of covering lamp stands, candlesticks, cornice-poles and rings of stove-dried wood with papier mâché and ornamenting in the usual manner. Papier mâché has been combined with metal as well as wood, and Mr. Benjamin Cook, junior, took out two patents in 1842 and 1843, respectively, for casing the pillars and other parts of metal bedsteads, cornice poles, cornices, finger-plates, trays, chimney-pieces, and a host of other articles, with papier mâché. A thin sheet of metal is first fashioned to the requisite shape, and then covered with layers of paper or pulp.

The applications of the material are too numerous to enter upon fully, and we can only refer to a few more particularly connected with the town. A patent was granted in 1778 to Clay, for making buttons of papier mâché. The blanks were cut from sheets and turned in a lathe. Shanks were to be fixed by boring a hole and passing a loop of wire through, the ends of which were bent down flat and covered with a layer of paper. In 1786 he had another patent for making buttons from fibrous pulp sheets, by pressing them to shape between dies when in a heated condition, "as the common horn buttons are made." This patent was granted on the 9th of November ; and on the fourteenth of the following month Obadiah Westwood, button maker, obtained protection for the very same method, except that instead of warming the blanks he warmed the dies. The title of this patent mentions "trays, caddies, boxes, bottle stands, inkstands, picture frames, mouldings, cornices, &c., &c.," but the specification is confined to button making. The language used to describe the ornamentation of buttons with pearl shell and other materials bears a very suspicious resemblance to Clay's patent of 1778. To take a rather long leap, Messrs. Souter and Worton patented in 1851 a method of making buttons and beads in a fly press or by rolling strips of the material through rolls on which suitable depressions are sunk. The buttons so made are finished in a fly press, and may be ornamented with pearl. They also describe a method of strengthening such articles by a layer of metal. In 1855 Mr. Fred. Walton, of Wolverhampton, proposed to make papier mâché trays with glass bottoms—if such a mode of expression be permitted—the rim being the only part where papier mâché is used. The beautifully decorated trays which were made a good many years ago were, in fact, only fit to be looked at. The present writer has in his possession the set of three trays representing "Spring," "Autumn," and "Winter," shown by Mr. Walton at the exhibition held at Birmingham in 1849, in connection with the visit of the British Association to the town. They are exquisitely painted, and are far too good to be used. One other application of the material due to a townsman may be noticed, and that is the proposal of Mr. Theodore Hyla Jennens to make calico printing rollers with a core of papier mâché on a mandril of iron, the copper shell being subsequently drawn over the core to compress it. This was patented in 1857.

Although candour has compelled us to point out that one or two inventions in connection with the trade originated elsewhere, the credit of bringing them into practical working must, undoubtedly, be awarded to Birmingham and the district. Tested by the number of manufacturers this branch of industry is declining, and there are now only ten houses engaged in the trade in Birmingham, and two in Wolverhampton. These figures show a reduction from those given in Mr. Aitken's article already referred to.

## CHAPTER X.

### THE BUCKLE TRADE.

Rise and Progress of the Trade — Rolling Buckles—An Encyclopædic Patent—Boulton's Spring Latchets—The Buckle Trade a Century Ago—Its Decay.

The laborious researches of the astronomer have enabled him to predict with unerring certainty the time when the most erratic of comets will again visit our firmament ; but who has determined the "periodicity" of a fashion? It has been said that there are signs of the re-appearance of the buckle, and we believe that a few feminine enthusiasts have hunted up some of those elegant ornaments from the collections of knick-knacks which belonged to their great-grandmothers. But anything like a general revival ot the fashion seems to be far distant. It had a very long run during the 17th and 18th centuries, but it became practically extinct about 1800. In 1791 the trade was in a very depressed condition, as we know from a petition which was presented to the Prince of Wales in that year, setting forth that the change of fashion had caused great distress amongst the 20,000 persons who were at that time employed in the manufacture of those articles. A slight revival took place in consequence of the renewed patronage of Royalty, but the decrees of fashion proved stronger than even those of a prince of the blood, and the buckle gradually disappeared. The patent records exhibit the decadence of the trade in a remarkable way. From 1779 to 1790 no less than 27 patents for buckles were taken out, or about 5 per cent, of the whole number granted during that period. Ten of these originated in Birmingham. Only three more buckle patents were issued in the remaining years of the century.

It should be understood that the buckles worn by our forefathers were not merely ornamental, like those which one now sees on ladies' shoes—they were in stage parlance "practicable"—that is, they really served to fasten together the straps of the shoe. The mechanism was however not usually visible, being concealed by a piece of leather which filled the ornamental frame or rim. The various inventions which we shall describe have reference to methods of producing the ornamental part, and also to the construction of the "chape" as it was called, that is, the mechanism for holding the shoe strap.

The first Birmingham patent, dated 1772, is in the name of William Tutin, a well-known buckle maker of the last century. His process consisted in silvering buckles and blueing them by means of varnish coloured with Prussian blue. The same method applied to buckles of "tin, pewter, spelter, or any white metal except iron or steel" was patented in 1785 by John Cooper, japanner. Tutin, who enjoyed a considerable reputation in his day, carried on business at 12, Coleshill Street. He was the inventor of a particular kind of alloy used for making buckles, called "Tutania," the origin of which would probably have been lost but for a casual reference in one of the lucubrations of the poet Freeth, reprinted in Mr. Langford's book. The inventor is thus alluded to—

All to one good soul must truckle,
He that does the rest eclipse,
Makes a song and forms a buckle
Whilst a pipe's between his lips.

Now farewell to vain disputing,
Of the evening make the most,
FRIENDSHIP, FREEDOM, TRADE, and TUTIN,
Round the Board shall be the Toast.

The poet adds in a note that Tutin was "a friendly cheerful companion, and exceedingly fond of a pipe." The composition of the metal is recorded in Richardson's "Chemical Principles of the Metallic Arts" (1790), a book which is full of interesting particulars relating to the trades of the town at that period. It consisted of 8oz. of brass, 2lb. regulus of antimony, and 7oz. of tin.

The next patent is one granted to William Bell, jeweller

and toymaker, in 1779, for a method of forming buckles and other objects by passing the metal through rolls upon which a device of the article required is sunk. The idea of producing ornamental articles by rolling instead of stamping has been patented perhaps fifty times, but this is not the first, as it was included in the patent granted in 1769 to Richard Ford, an ingenious mechanic of the town, who built Hockley Abbey. We shall have something to say of Ford in a subsequent article.

If the invention disclosed in the specification of Daniel Winwood, buckle chape maker, dated May 28, 1781, was really new at the time, then he must be regarded as the inventor of what is known as "joint-wire," so largely used in the jewellery trades. One part of Winwood's invention consists in making the hinges of buckles by soldering short pieces of tube of very small bore—or, in fact, hollow wire—through which a pin is afterwards passed. This is the method adopted for fashioning the hinges of lockets, brooch-pins, and similar articles. There is some reason to suppose that this ingenious device is of French origin, the hollow wire employed being known amongst workmen as *charnière*—the ordinary French word for hinge. According to another part of the specification the hinge is formed by simply turning over the metal. In a patent with a similar title applied for in the previous year, but never completed, Winwood is described as an ironmaster, which, in the phraseology of the day, meant no more than ironfounder. In all probability he was identical with the Winwood referred to in Mr. Thomas Hadley's evidence before a committee of the House of Commons in 1799, as having gone to Vienna with his two sons to superintend a button and buckle manufactory there, to the great injury of the Birmingham trade. In the "Directory" of 1770 a Daniel Winwood appears amongst the toy-makers, as carrying on business at 35, Newhall Street.

Although not strictly a Birmingham patent, we must not pass over that granted to William Playfair in 1785, at which time he was residing in London. We shall have something to say of the inventor on a subsequent occasion, meanwhile it will be sufficient to mention that he was a brother of the celebrated Professor Playfair, and that he was employed at Soho for some years. His name appears as a witness to Watt's signature in the copying-machine patent. Playfair proposed to prepare the metal for making buckles by rolling it in the form of rods through rolls having grooves of the required section, either engraved with ornaments or plain, according to circumstances. The ornament may also be impressed by stamping or swaging.

For plain buckles the metal is drawn through dies. A sufficient length is then cut off, bent to the outline of the buckle, and soldered. Buckles and other articles are plated by interposing a thin layer of fusible silver-alloy between the surface to be coated and the plate of silver, for the purpose of facilitating the junction of the two.

Mr. James Alston's patent of 1785 is a perfect encyclopædia on the subject of buckle-making. Not only does it contain a description of his improvements, but in addition it gives a minute detail of the mode of manufacture then followed. The principal object of the invention is to cover the edges and bottoms of buckles with a durable coating of tin, silver, or gold, and to obviate the want of durability of buckles plated on tin. It appears from the specification that iron buckles became unsaleable from rust, and those plated on copper or brass were in almost total disuse in the home trade from their edges being seen. The inventor also proposed to shape the gold and silver plates previous to attaching them to the buckle by rolling instead of stamping. The James Alston here mentioned was probably an ancestor of the present Mr. W. C. Alston, of Elmdon Hall. We have already referred to the Alston family in connection with the early manufacture of oil of vitriol in the town; but we are not able to state with precision the relationship which existed between the various persons of that name.

In the following year, Thomas Cheston, plater, obtained a patent for making plated spring buckles, but the specification is very obscure; and in 1786 John Rose, button maker, applied for protection for a mode of ornamenting steel buckles, but, as the patent never arrived at maturity, we have nothing but the bare title of the invention. In 1790 Richard Law, plater, patented a buckle fitted with slides, so that the ornamental part might be removed and another substituted at the pleasure of the wearer. The inventor also describes two or three methods of securing the straps by hooks placed upon the frame, or by pins on a cross bar or "bridge" as a substitute for the ordinary chape. A very similar idea was patented two years earlier by Joseph Jukes, a plater in the town. These contrivances received a further development a little later on, as we shall presently see.

Samuel Hands's patent of 1790 was referred to in the previous article. It consists of a method of inlaying buckles "made of leather or paper," with ornaments of gold, silver, or other metals, glass, pearl, ivory, &c., by cutting, drilling, chasing, or pressing. This individual formed one of the deputation to the Prince of Wales mentioned above.

As regards the buckle proper, invention seems by this date to have come to an end, but an attempt was made in 1792 to introduce a sort of spring catch by James Smith, Boulton's partner at Soho, where the buckle business was carried on by the firm of "Boulton and Smith." It has the appearance of an ordinary buckle, but underneath there is a spring, at the extremities of which are hooks, the points of which are set inwards. The buckle is fixed by pressing it on the shoe, and when the pressure is released the spring forces the points into the straps and holds them together. Mr. Jaffray in his "Hints for a History of Birmingham," which appeared in these pages many years ago, states that this fastening never met with much favour, and that it died out about 1804. The present writer has in his possession some of Boulton and Smith's original buckles, still attached to the cards as they left the factory. The notion of a hook fastening seems to have caught the public taste, if we may judge from the fact that another patent was taken out in 1794, by William Bell, a Walsall buckle maker, for a very similar, if not identical contrivance. It only differs from Smith's in having a ring underneath through which the shoe strap passes, so that when it requires to be tightened it is pulled in a direction away from the centre of the foot. A very elaborate drawing is attached to the specification, showing various forms of the fastening, and also the mode in which it is attached to the shoe. We should judge it to have been a better thing than Smith's. The inventor was probably identical with the William Bell previously mentioned, but he had, as will be seen, removed to Walsall. Bell's patent was dated May 17 ; and in September of the same year we find a patent by one Benjamin Adams, a Birmingham buckle maker, for "shoe catches with spring straps, to be used with buckle rings of every kind and description." An ordinary buckle frame is fitted with two or three sharp hooks at each side, pointing inwards. The shoe straps are tightened by being pulled towards the centre, the hooks taking hold of the leather and preventing them from slipping back. An ornamental centre, made slightly elastic, is "sprung" into the buckle frame and conceals the ends of the shoe straps. According to another modification the ornamental part is hinged and fastened with a catch. Many years afterwards, in 1811 a London goldsmith named Davies took out a patent for a somewhat similar fastening, in which the straps are fastened by being hitched on to a stud fixed in a bar across the centre of the buckle. The straps are prevented from coming off the studs by a hinged piece, which is folded down and kept close by a catch.

Amongst the various patterns made at Soho was one in which the chape was hinged on a bar to give increased facility for inserting the straps of the shoe. It is difficult to say who was the inventor of this particular form, as there are two patents for it, one in the name of John Stedman, dated June 30, 1788, and another in that of Thomas Freeth, dated July 2, 1789. Both patentees were resident in London. A comparison of one of the actual buckles manufactured at Soho, with the drawings attached to the respective specifications, does not enable us to form any certain opinion on the point.

The only other Birmingham patents which remain do not relate to shoe buckles, but they may perhaps be noticed. J. H. Rodgers in 1839 invented a form of buckle in which a lever is substituted for the tongue usually employed. The lever does not penetrate the strap but simply exerts a powerful pressure upon it, and prevents it from slipping. This has been applied to braces. In 1843 Thomas Simpson brought out a buckle with a detachable tongue, permitting the ready attachment of a strap. In 1846, Mr. Charles Rowley, the inventor of many ingenious contrivances in his trade, patented machines for bending wire for buckles, and for stamping buckle tongues.

Allusion has already been made to the fact that there were 20,000 persons dependent upon the buckle trade of Birmingham and the neighbourhood towards the end of the last century. It may be assumed that these figures merely referred to the manufacture of dress buckles. The materials are not at hand for instituting an accurate comparison, but it will be instructive to state that the current edition of the Post Office Directory gives the number of buckle makers in Birmingham as 4, and buckle tongue makers, 2. In the entire county of Stafford there appear to be, buckle makers, 25; buckle platers, 2; buckle tongue makers, 3; buckle tongue platers, 1; or 37 houses in all in Birmingham and Staffordshire. According to Sketchley and Adams's "Directory," for 1770, so frequently referred to in the course of these articles, there were at that time 58 buckle makers in Birmingham; 127 in Wolverhampton; 283 in Walsall; 1 in Willenhall; and 68 in Bilston; or 537 in all. We do not desire to make more of these figures than the facts actually warrant—for instance, many of the establishments in the Black Country must have been exceedingly small—but they illustrate in a curious way the decay of a great trade.

# CHAPTER XI.

## THE BUTTON TRADE.

PATENTS FOR BUTTONS—EARLIEST PATENT FOR BUTTONS—STAMPED AND PRESSED BUTTONS—PAPIER-MACHE AND PULP BUTTONS—CAST BUTTONS—BUTTON SHANKS—WHITMORE'S PATENT—RALPH HEATON'S BUTTON MACHINES—HICKMAN'S PATENT.

Birmingham and buttons have long been closely associated, and if ever there was a strictly "local" manufacture it was the metal button trade. The patent records show this in a remarkable way, and out of the 21 grants for improvements in button-making between 1770 and 1800, no less than 19 came from Birmingham. If we take a longer period the supremacy of Birmingham is still maintained, though not to such a striking extent. From the year 1631 to 1852, 112 patents were issued, 70 of which originated in the town, and many of the exceptions are not improvements in making buttons, but mere theoretical suggestions for the application of certain materials. A glance at the list of patents in chronological order reveals the fact that a period of great inventive activity was followed by one during which the subject seems to have been entirely forgotten. For instance, between 1785 and 1787 no less than 11 patents were issued, all from Birmingham, except one for porcelain buttons. There were two in 1790, one in 1794, one in 1800, then nothing until 1813, followed by a long interval until 1824, when there was again one. In 1824 the trade seems to have revived, and one patent was the result; in 1825 there were two, in 1827 and 1829 one each, and after 1831 several were issued every year.

A century and a half ago the button trade of the town was already of great importance, but the earliest Birmingham button patent was not taken out until 1770. Before describing it we will quote the account prefixed to the list of button makers in the old "Directory" which has been so frequently alluded to. The compiler says:—"This branch is very extensive, and is distinguished under the following heads—viz., gilt, plated, silvered, lacquered, and pinchbeck, the beautiful new manufacture platina, inlaid, glass, horn, ivory, and pearl; metal buttons such as Bath, hard and soft, white, &c. There is likewise made link buttons, in most of the above metals, as well as of paste, stones, &c., in short the vast variety of sorts in both branches is really amazing, and we may with truth aver that this is the cheapest market in the world for these articles." It should be explained that the "platina" above mentioned was not the metal platinum, but an alloy consisting of 8 parts brass and 5 of spelter.

The patent just alluded to as being the first was granted in 1770 to John Smith, jeweller, who, according to the "Directory," was also a "Philligree worker," and carried on business at No. 14, Newhall Street. It has reference to a method of stamping the cups for gold or silver buttons which are to be set with stones, and cutting out the "cramps" on the edge of the cup which, when turned over, serve to hold the stone securely in position. If it be true, as hinted in the title, that such buttons had not hitherto been made out of a flat disc, then the invention would seem to mark a distinct step in the trade. The specification is accompanied by a drawing of the dies, and the process is described with a certain amount of detail. In the title mention is made of "a new letter link to the buttons to distinguish them from the buttons of other makers," but there is no description of this part of the invention.

In the interests of art one cannot but regret that Isaac Whitehouse, enameller, of 36, Bull Street, should have so far forgotten what was due to his craft as to propose to ornament "buttons, studs, bracelets, necklaces, lockets, rings, watch chains, boxes, and trinkets," by pasting silk upon paper and covering it with glass fixed in mounts or settings varying with the nature of the particular article. Whitehouse's patent is dated August 31, 1774.

Although buttons are mentioned in the title of the patent by William Collins, button and toy maker, granted in 1778, it does not refer to the actual manufacture of those articles, but only to a mode of burnishing plated metal suitable for such purposes by passing it between rolls, one of which is caused to move faster than the other. The patentee also appears to claim the manufacture of buttons from sheets of metal already gilt, instead of gilding each one separately. This Collins was the inventor of the alloy which is alleged to have been substantially the same as Muntz's metal, as already stated in the fourth article of this series. He removed to London towards the latter end of the century, and became agent to a copper company, in which capacity he gave evidence before the Parliamentary Committee on the causes of the high price of that metal in the year 1799.

Passing over the proposal of William Bell (1779) to produce buttons by rolling from bars of metal, we come to James Alston's patent, dated 1785, for making sheet iron buttons by stamping red-hot blanks in a die, either engraved or plain. The fraze or selvedge having been cut off in an "edging lathe," the shanks are to be attached by soldering, screwing, or riveting. Minute directions are given for tinning, silvering, or gilding buttons of this kind, and they may be inlaid with other metals by a blow with a stamp. He also describes a mode of gilding and silvering cast-iron buttons. This patent was referred to last week as containing much information on the manufacture of buckles. The same observation applies to the subject of this article. It is very long, and, as the inventor himself remarks, "it seems impossible to enumerate every means of carrying out the invention." There is, however, one feature which should not be overlooked, and that is a mode of covering the face of iron buttons with a thin plate of brass or "Bath metal," which is folded underneath by suitable dies. In 1787 Alston took out another patent for tinning cast and wrought iron buttons. It is not so long as the previous one, but it contains many details which are not *ad rem*. We learn incidentally that "tinned iron"—the tin-plate of our day—was then "commonly called sheet tin."

It must be admitted that many of the inventions now being dealt with do not display any striking amount of originality, as they are merely applications of new materials. Clay's and Westwood's patents for papier mâché and pulp buttons were alluded to in a former article. John Eginton's patent of 1786 merely consists in substituting wood for the papier mâché used by Clay. He proposes to use either solid wood, a material made by gluing veneers together, taking care to cross the grain, or a composition consisting of sawdust, chaff, hemp, straw, &c., cemented together by glue. The mode of fastening the shanks is also similar to that described by Clay, which consists in passing the ends of the wire through a hole in the button, and bending them down flat. A piece of material is glued on the face to hide the ends of the shank and keep it secure. This Eginton, who describes himself as an engraver, was in all probability related to Francis Eginton, the well-known stained-glass artist, but in what way we have not been able to ascertain. In 1790 Clay suggested the use of stone or slate as a material for buttons, the shanks of which were attached in the same manner as that just described, or they might be riveted through, the end of the shank forming an ornamental stud in the centre.

Cast buttons have been known for a very long period, and a patent was granted in 1683 to Richard Maundrell and John Williams, who state that they were "the first inventors of the mistery of casting and makeing hollow pewter or block tinn buttons." The first Birmingham patent relating to this branch of the trade was however only issued in 1786 to John Francis, founder, whom we take to be the partner of Richard Dearman, of the Eagle Foundry, Broad Street. The moulds are formed from patterns of lead or other material, and the buttons after having been annealed for twelve hours are scoured, turned, and cleaned ready for japanning, painting, or otherwise finishing. In the same year Joseph Ashton, of Bordesley, iron founder, took out a patent for "casting and making buttons and button shanks." The shanks are cast at a separate operation and then laid in the sand and the buttons cast round them, a depression being made in the stem of the shank to cause the metal to hold better. Patents were granted to a Joseph Ashton in 1769 and 1771 respectively, for making cast nails. In one of them he is described as a "caster," and we think that all three may with certainty be identified as being the Ashton of "Francis and Ashton, nail factors and manufacturers, of Moat House," in the "Directory" of 1770.

There is one feature in the patent of Joseph Rabone, button maker—granted in 1787—which is worth calling attention to, as it has probably been in uninterrupted use from that day to this. The recess for the shank of bone and ivory buttons is slightly undercut, whilst the disc to which the shank is attached is made concave. It is then introduced into the recess and flattened, which enlarges it, causing it to spread out to the diameter of the bottom of the recess. This is an exceedingly trifling matter, but it is very ingenious, and the name of the presumed first inventor is worthy of being put on record. The specification contains many details of the processes used in those days for ornamenting the buttons then in fashion. "Some," he observes, "are to be wrought in scoring engines, which are now in daily use by myself and many other gilt and plated button makers." This was no doubt the "rose engine," as it is now called, by the use of which very beautiful effects in "engine turning" were produced.

Considering the enormous demand for buttons, it is not suprising that plans of making the shanks should have been devised. We find accordingly that in 1790 William Whitmore, tool-maker, took out a patent for a method of rolling button shanks from fillets of strips of metals, and afterwards cutting them out by a punch and bed fitted into a press or stamp. He was also the inventor of an improved mashing-machine, and of a method of preventing the wear of the bearings of weighing machines by enclosing them in a box of oil. Both of these contrivances are referred to in Bisset's "Poetic Survey." Whether Whitmore's plan was a commercial success or not we are unable to say, but it was eclipsed by Ralph Heaton's machine which was patented in 1794. The inventions of the Heaton family are numerous and important, and demand a special article, but we must refer to them under their various subjects as we go along. It would be impossible to convey any idea of the *modus operandi* of the button shank machine on account of its complicated construction, and all we can say is that it is entirely self-acting, the wire from which the shanks are fashioned being seized by rollers, cut, bent, and twisted until it has been made to assume the required form. Apart from the special subject with which we are now dealing, the machine is of great interest as being probably the first example of a machine in which the material is subjected to a series of independent operations succeeding each other at regulated intervals of time. We are familiar enough now with composite pieces of mechanism of this kind—such as the machines for making pins, hooks and eyes, wire cards for carding cotton, and so on, but the button-shank engine preceded these by many years. As a proof of Ralph Heaton's sound workmanship and correct principles of construction, we are able to say that some of these identical machines are still at work at Messrs. Heaton and Dugard's mill in Shadwell Street. Heaton's place of business was in Slaney Street, where he acquired a considerable reputation as a manufacturer of rose-engines, and of the various tools used in the Birmingham trades. He was also the inventor of a machine for drilling four holes at once in buttons, which is probably that described in Hebert's "Engineer's and Mechanic's Encyclopedia;" and in 1790 he contrived a press for cutting out and punching the holes in button cards at one operation. This was made for John Tookey, who died soon after, leaving it to his brother William Tookey, wire drawer, of Dale End. The latter continued to be the only maker of button cards in the town for many years. It may seem a singular transition from button cards to biscuits, but a machine on a similar principle is used at the Victualling Yard, Gosport, for cutting out and punching biscuits for the Royal Navy. Ralph Heaton died in 1832, and is buried in St. Paul's churchyard. The present Mr. Alderman Heaton is his grandson.

Before concluding the history of the trade down to the year 1800, we must mention the patent granted in 1787 to Robert Hickman, of the parish of Edgbaston, button maker, for "uniting or amalgamating with, or by means of tin, or tin and lead mixed, gilt and plated button shells, both coloured, plain, and figured, with bottoms of copper, brass, iron, mixed and compound metals." The nature of the process is sufficiently indicated by the foregoing extract from the title of the patent. This Hickman was in all probability the person of that name referred to by Mr. Thomas Hadley, in his evidence before a committee of the House of Commons in 1799, as having been induced to go abroad to superintend a button manufactory at Vienna. John Rose's patent of 1787 related to steel buttons, but no details of the process can be given, as the patent was never completed. In 1800 Joseph Barnett, merchant, took out a patent for a button with two shanks, placed opposite to each other. The button was to be sewn on with the shanks in the axis of the button-hole. This was said to give a more flexible attachment, and to cause the button to sit better to the garment.

Towards the close of the last century the metal button trade was in a very depressed condition, owing partly to the very high price of copper, and also to the pressure of foreign competition. In his evidence before the Parliamentary committee above alluded to, Matthew Boulton stated that he had some short time previously given up the trade as not being sufficiently remunerative. The persons to whom the business had been handed over were equally unsuccessful, and they also had relinquished it. Some very interesting particulars of the decline of the Birmingham button trade were given by Mr. Thomas Osler, who attended as one of the deputation from the Chamber of Commerce, in his evidence before a Parliamentary committee on the emigration of artisans in the year 1824. He spoke of a certain "fancy-cut white metal button, cut by an engine, a showy and cheap thing," in the manufacture of which forty or fifty houses were once engaged. A manufacturer who thoroughly understood the trade was detained in France during the war, and, under the patronage of the French Government, he commenced to make these buttons. The result was that the Birmingham trade was ruined, and "France has now for some time supplied all Europe with that article, and distant markets." He also referred to the "Bath metal drilled shank button," which was cast whole, stamped, and the shank drilled. Twenty thousand gross per week were once made in Birmingham, but the high price of labour and the duty on copper caused the trade to decline.

## CHAPTER XII.

### THE BUTTON TRADE—(CONTINUED).

DOUBLE-SHANK BUTTONS—DR. CHURCH'S PATENTS—HORN BUTTONS — INGRAM'S PATENT — ROWLEY'S PATENTS—PROSSER'S AGATE BUTTONS—DRY CLAY BUTTONS AND TILES—FRENCH COMPETITION.

As we showed in the last article the button trade was apparently suffering from a collapse during the first quarter of the present century, in so far, at least, as the number of patents granted may be regarded as evidence of the fact. Barnett's double-shanked button, patented in 1800, has already been referred to, and the next on the list is the patent granted to James and William Henry Deykin in 1824, for making uniform and livery buttons, and consists rather in a change in the order of the various processes of stamping, annealing, attaching the shanks, and pickling, than in any novelty in the method of manufacture.

Amongst the numerous subjects to which the ingenious Dr. Church—who is still remembered by many Birmingham men—directed his attention was that of buttons. His first patent was granted in 1827, and is taken out in the name of Thomas Tyndall. Like all his inventions it is very complicated, but the object is to produce a button with a peculiar shank formed by raising a convexity in the centre of a metallic disc, and piercing holes through the raised portion in such a way as to form a straight passage for the needle and thread, parallel to the face of the button. It was, in fact, an attempt to secure the same amount of flexibility as that belonging to buttons with a projecting tuft of fibrous material, or "flexible shank" as it is called. But the idea did not originate with Dr. Church, for the self-same principle is laid down in De la Mayne's patent for porcelain buttons, granted in 1785. In 1829 Dr. Church patented certain improvements upon his former method, one of which was making the shanks separate, by punching out blanks in the shape of a cross, which were then dished and the arms turned up into a wire like form. They are fixed in the button by folding over the upper shell. Both of these machines include also arrangements for covering buttons with cloth.

As the manufacture of flexible-shank covered buttons will form the subject of a separate article we pass over for the present Aingworth's patent of 1831, and proceed to notice Mr. T. W. Ingram's patent, granted in 1832, for a press with a peculiar arrangement of toggles for pressing horn or similar buttons. The machine is well adapted for giving the slow but powerful pressure required in fashioning such articles, so much so indeed, that a very close imitation of it was patented many years afterwards by one William Tudor Mabley, of West Lambrook, Somersetshire. Although the John Holmes to whom a patent for button shanks was granted in 1833 is described as of Worcester, we believe him to be identical with a person of the same name who subsequently resided in Birmingham, where he took out several patents. His shanks were raised out of the solid, and he also describes various methods of attaching them to the button, one of which is the very old and frequently patented device of undercutting the recess in the back of the button. In 1835 Humphrey Jefferies, the inventor of the linen or sewn-through button, proposed to form the shanks of brace buttons by raising a swell on the underside and punching a hole in it, the edges of the hole being turned under so as to form a perfectly smooth surface which will not cut the thread.

Horn buttons were made in the town more than a century ago, but the first attempt to give an artistic character to the manufacture dates from 1837, when T. W. Ingram took out a patent for ornamenting the surface by the use of engraved dies. The metal employed was brass, pieces of which, suitably engraved, were fitted into cavities cut in steel or other plates. The dies were engraved flat, and subsequently bent to any required curvature. The patent included a method of attaching flexible shanks by sinking a suitable recess in the back of the button. Ordinary shanks were also attached when desired. The manufacture was taken up by Mr. T. Cox, and some of the buttons made by him were very effective, but the method employed for securing the flexible shanks was not successful, and the buttons required trimming after leaving the dies, on account of the spreading of the material beyond the edge. To make a better article was the object of Joseph Parkes's patent of 1840, and he proposed to cover the fronts of buttons with very thin horn, which, when softened by heat, has sufficient flexibility to allow of being treated somewhat like a textile fabric. In the next year Mr. Thomas Harris, of Hales Owen, patented a further improvement, which consisted in fixing the flexible shanks at a separate operation when the horn is not so highly heated. The buttons were inlaid with devices in metal, much in the same way as that employed for fixing the shanks, and they were gilt or silvered by attaching leaf metal with cement. A method of engraving the dies is described, as also an arrangement for preventing the spreading of the horn beyond the edge of the dies, so that the buttons do not require to be trimmed. The manufacture of horn buttons is still carried on at Hales Owen by Mr. W. J. Harris.

The button patented by Henry Gibbs in 1838 may be best described by comparing it to an ordinary "collar stud," the holes for attaching it to the garment being pierced round the edge of the larger disc. Rowley and Wakefield's patent, of 1840, is rather an improvement in stamping and piercing in general than a button patent, as it does not contain any improvement in the actual mode of making buttons. It is capable of cutting out and punching holes in blanks at one operation, the piercing tools sliding within the punch and coming into operation after the disc has been cut out. It includes a method of working stamps and presses by power, and an automatic apparatus for feeding the strips. There is also a very ingenious machine for rolling and rounding the edges of button blanks, consisting of a roll working inside a steel cylinder. The blanks are fed in at an aperture in the cylinder, and are carried round by the action of the roll and discharged through a corresponding aperture at another part of the circumference. Except that the action is continuous, instead of intermittent, the principle is similar to that of the old machine, in which a straight bar of steel is caused to vibrate in a direction parallel to a similar fixed bar, the blanks being placed between the two. (An apparatus of this kind, with automatic arrangements for feeding the blanks, was patented by Mr. George Twigg, in 1852.) In 1841 Messrs. Chambers, Franklin, and Rowley patented a particular form of shank raised out of a flat disc with the holes at right angles, as in Dr. Church's patent of 1824. The specification also contains a claim for the application of vegetable ivory as a material for making buttons, but we believe that it did not come into use for some time afterwards. In the same year Mr. Benjamin Aingworth, whose specifications abound in suggestive detail, obtained protection for button shanks of various kinds, and for modes of fixing them on buttons of porcelain, glass, and other materials. He also has a shank in which the holes are at right angles, like that alluded to above. There have been many contrivances for rounding the edges of the holes of perforated buttons, so as to prevent the thread from being cut, and Messrs. Rowley and Turner (1842) proposed to effect that object by gathering the metal round the edges, so that the button appears as if rings had been let in. In the same year William Bridges patented a button with a detachable shank, to be used where it is desired to wear the same garment with different sets of buttons, as was the fashion at one time. In 1844 we find two patents for pulp buttons, one by Mr. Benjamin Aingworth and the other by Mr. William Sheldon. The latter professes to give such buttons the appearance of those made from horn, which is also a feature in Mr. William Elliott's patent of the following year. Messrs. Newey and Newman's patent of 1844 included a peculiarly shaped button, the circumference of which was formed of a hollow ring, the two discs forming the shell being turned up and then pressed together. This patent was worked by Mr. Elliott, under a license from the patentees. Mr. Rowley, in his patent of 1846, claims amongst other things an improved four-hole button and a method of applying shanks to horn and pearl buttons. Lutwyche's improvements in porcelain buttons (1846) consist in fixing shanks by undercutting the buttons before being completely fired. The specification of Mr. Samuel Kenrick's patent, of 1847, is a sort of small treatise on the art of what is known as "odd-side moulding," and contains minute directions for preparing moulds for buttons, nails, and other articles. A particularly ingenious contrivance is described for placing button shanks in the mould in readiness for the metal to be run in to form the body of the button. In the same year Mr. Arthur Boyle patented a mode of making buttons by bending wire into a form somewhat resembling the figure 8 flattened, and afterwards finishing them in dies.

To fully describe Mr. Rowley's patents of 1848, 1849, and 1850 would require an entire article, but we can only briefly say that they are full of the most ingenious devices for stamping and punching buttons and similar objects, such as weavers' mails, &c. One of them includes a method of cutting ivory buttons in facets, and in that of 1850 will be found a method of attaching buttons to waistcoats and shirt studs to shirts. In 1849 Charles Iles patented a plastic composition mixed with shreds of coloured fabric, such as silk, for the purpose of producing buttons having a veined or marbled appearance. A clever notion is embodied in Newey and Newman's patent of 1849, consisting of ornamental buttons having the appearance of a number of gems in metal settings. A disc of glass—china may also be employed—is, first of all, embossed with any suitable pattern, and a metal plate, perforated to correspond with the projecting parts of the design, is then dropped on to it. (A very similar combination was subsequently patented by Messrs. Souter and Worton in 1851.) They also proposed to produce ornamental effects by superposing layers of differently coloured glass, and cutting through the outer layer—a process which, however, was very well understood, and frequently practised, by the ancients. The patent included, in addition, a method of preparing metal sheets for making buttons by passing them through rolls engraved with figures of the button required.

Amongst what may be termed the miscellaneous processes of button-making there is one which was invented by a Birmingham man, although it was never carried out here, nor, indeed, in this country at all to any great extent, except for a short time. We refer to Mr. Richard Prosser's "agate" buttons, as they were, not unhappily, called, for which a patent was granted in 1840. The process, which was exceedingly simple, consisted in pressing a finely-powdered mixture of clay, felspar, flint, &c.—the proportions of which vary according to circumstances—between dies of the shape of the button required. A single pressure only was requisite, and the buttons were then fired and glazed in the usual manner of treating porcelain articles. The process was taken up by Mr. Minton, of Stoke-on-Trent, and at the beginning of the year 1842 a make of 5,000 gross per week was found insufficient to supply the demand. For a few years the process was successfully carried on at Stoke, and shortly afterwards the manufacture was commenced in France. M. Bapterosses took out several patents for improved machinery by which he was enabled to make no less than 500 buttons at a single operation of the press, and, in consequence of the low price of labour in France, the "agate" buttons could be sold here for less than it cost to sew them on to the cards at Stoke. This, of course, ruined the home trade, and they have ceased to be made here for many years past. There is, however, a very large establishment at Briare, about forty miles from Orleans, where they are manufactured in enormous quantities. Mr. Prosser's patent included also a claim for a new form of button with two holes connected by a channel for the thread to lie in, which had the effect of protecting it from wear. At the suggestion of Mr. Blashfield the process was applied to the making of tesseræ and tiles, and the foundation of a most important industry was thus laid. An early use of the patent tesselated pavement in Birmingham was for the steps of Mr. Prosser's house, No. 18, Broad Street. We have ventured to digress thus far from the subject in hand to place on record a few facts, which seem in danger of being forgotten, relating to the origin of the "dry clay" process of making tiles and buttons.

## CHAPTER XIII.

### THE BUTTON TRADE—(CONCLUDED).

CLOTH BUTTONS — SANDERS' PATENT — ELLIOTT'S PATENT AND LITIGATION — JEFFERIES' LINEN BUTTON — EXTENT AND INGENUITY OF BUTTON PATENTS — VALUABLE SPECIFICATIONS.

NO single invention has exercised a more important influence on the button trade than the method of making covered buttons by machinery. It dates from 1813, in which year a patent was granted to Mr. Benjamin Sanders, a native of Worcester, where he was born about the year 1761. He was educated at the Cathedral School of that city, with the intention, it is believed, of taking holy orders, but when quite a young man he removed to London, and commenced business as a merchant. Subsequently he went to America, and settled at Long Island, New York, where he started a tannery. A sharp attack of yellow fever—Long Island was then little better than a swamp—caused his return to Europe, and in 1799 he arrived at Copenhagen, where he continued to follow the business of a merchant. He amassed considerable property, but sustained very severe losses in the bombardments of 1801 and 1807. Collecting the wreck of his fortune he came to England in the last-named year to prosecute his claims against the Government for compensation. How he came to turn his attention to button-making at an age when most men begin to think of retiring is not known, but he was endowed with remarkable versatility, and it is stated that some years before he had invented an improved form of joint for bedsteads, besides being the first to suggest the use of silk nap as a substitute for the beaver then universally used for hats. Like all inventors he had to encounter much prejudice, but he was a man of indomitable energy, and he had, somehow or another, succeeded in making powerful friends, amongst whom were the Dukes of York and Sussex, and General Grosvenor. By this assistance he was enabled to push his invention into notice. "Sanders's buttons" became the fashion, and tailors began to receive orders from their customers that no others were to be used. It appears from the London "Directory" that he began to manufacture the new buttons in 1817, at No. 25, Bedfordbury, where he remained until 1825. In that year he removed to Bromsgrove, where the trade is still carried on by his descendants. Sanders died in the early part of 1851, at the age of 90, and he lies buried in Bromsgrove churchyard. The family is a very old one in the Midland Counties. Benjamin Sanders's uncle was heir-at-law to Shenstone the poet. One of his grandsons now fills the office of High Sheriff of Herefordshire.*

Sanders' patent was for a method of making buttons covered with cloth by the use of a series of dies for consolidating the various parts, and he also describes a method of giving flexibility to the shank by interposing a spiral spring "between a small plate keyed or pinned through the shank behind the garment, and the garment itself." This contrivance may be regarded as the forerunner of the flexible shank in everyday use, which was patented by Mr. Sanders in 1825. A few years afterwards the manufacture of these buttons was commenced by Mr. Aston, of this town, but on an action for infringement being brought in 1831 Mr. Sanders was nonsuited, and he was unable to sustain a general claim to the manufacture of flexible shanked buttons. As the law stood at that time, a patent was voidable if it contained anything that was old, and Sanders had unfortunately described in his specification a particular kind of shank formed by interlacing loops of cord passing through the back of the button. It was proved on trial that this was not new, and for that reason, as well as on account of some technical defect in the "claim," the patent was held to be bad, although the specification was drawn by the late John Farly, who was *facile princeps* at the work. Aston offered to take a license at a nominal sum, and to keep up the credit of the patent, but this was finally refused by Sanders, who held that the patent was either his or it belonged to the public at large, and that he would not be a party to any arrangement. It was understood at the time that Sanders's case had no small effect in hastening the passing of Lord Brougham's Act of 1835, which enabled a patentee to cure such defects in their specifications by filing a "disclaimer."

Dr. Church's patents of 1827 and 1829, referred to in the last article, included a very elaborate self-acting machine for covering buttons — the only one of the kind that we are acquainted with — which contained mechanism for conveying the blanks to the

* There was a certain Mark Sanders, who carried on business as a metal button-maker in Jennens Row, Birmingham, in the early part of this century, and who is referred to in Rees's "Cyclopedia" (art. buttons) as the inventor or introducer of an apparatus for preventing the deleterious effects of the mercurial fumes evolved in the old process of gilding. He was no relation whatever to Benjamin Sanders.

dies, punches for cutting out the discs of cloth from a roll of that material, and contrivances for superposing the various parts of the button in their proper order. Mr. Benjamin Aingworth took out a patent in 1831 for buttons, the back of which was almost entirely covered with fabric, which served for attaching the button to the garment. There was no metal "collet," and no projecting tuft on the back, so that the button-hole was not worn so quickly. This patent was assigned to Messrs. Sanders, the consideration being an annuity of £300. per annum to Aingworth, who was then only a garret-master. The major portion of the patent was in reality due to Sanders and his son, but in consequence of the keen competition which then existed it was deemed prudent to purchase Aingworth's invention, even at a very high price. A patent was granted to Mr. John Aston, in 1834, for a peculiarly formed flexible shank, which did not protrude from the general face of the button and was found to sit much closer to the garment. In the collet there was an oblong hole, having the threads which act as the flexible shanks laid across the short axis of the hole. Samuel Burrell's patent of 1835 consisted in using a soft metal plug for connecting the front and back plates of the button. The plug was placed under the fabric constituting the flexible shank, and caused to spread by the pressure of a pair of dies.

It is tolerably safe to conclude that when much litigation arises about a patent it is both valuable and successful, and this was certainly the case with Mr. William Elliott's patent of 1837, which consisted in a method of making buttons covered with velvet or other silk fabrics having a pattern or design in the centre. It will be obvious from what has been already said that the methods previously in use were inapplicable to the manufacture of such articles, as the pressure of the dies would injure the face of such materials, and special means would be required to ensure the pattern always being exactly in the centre of the button. Mr. Elliott was compelled to embark in a costly and protracted litigation to protect his rights, first, in 1840, against Mr. Aston, and again in the year 1842 against Messrs. Turner. He was ultimately successful, but the case of Elliott *v.* Turner went on in various forms for more than four years, although the point at issue was almost microscopic. Messrs. Turner were working the patent under a license, according to which they were to pay a certain royalty. In the course of business they manufactured a quantity of "Italian twist dress buttons," and the question was whether such buttons came within the patent. The judge before whom the case was first tried appears to have taken a very narrow view of the words "soft or organzine silk" which occurred in the plaintiff's specification as defining the fabrics to which his invention applied. He held that unless the silk was "organzine" it was not within the patent, against the plaintiff's contention that soft silk, though not "organzine," was included in the specification. The case furnishes an interesting instance of the extreme difficulty which sometimes arises in construing specifications, and it took no less than *eight* of the most eminent judges on the Bench to decide that the original ruling was wrong, and that the case must be tried over again.

No further patent was granted for covered buttons until 1841, when Humphrey Jefferies invented a mode of making them by using a single metal shell having a hole for receiving the protruding flexible shank. In the same year Messrs. Chambers, Franklin, and Rowley patented a button with a collet of blued steel which was turned over the front, producing a button with a cloth centre and a steel edge. Mr. Aingworth also took out a patent for a mode of covering buttons, and for a flexible shank formed by working a twisted cord of silk or other thread crosswise through the back. Up to the time of Mr. Chatwin's patent of 1842 the covering fabric was held by folding the circumference of the front shell, in place of closing the covering fabric between the front and back shell, or collet. This was the essential feature of his patent, by which he effected a considerable economy in material. Mr. Wm. Bridges' patent of the same year included a plan for cementing the covering material to metal sheets from which the blanks were cut, suitable means being provided to ensure the central position of the pattern. The edge of the button was protected by turning up the collet much in the same way as that just described.

In 1843 Messrs. Aston and Elliott invented a method of economising material, the principle of which somewhat resembles that patented by Chatwin. They also proposed to apply india-rubber cement to the back surfaces of the covering fabrics of pressure-made buttons to prevent the patterns from being opened by the action of the dies. Mr. J. F. Empson's patent of 1844, which was subsequently assigned to Mr. John Aston, includes various improved modes of economising covering material, a new construction of flexible shank, and a method of making flexible shanks with braid, gimp, &c. It was subsequently found however that most of these improvements were old, and a "disclaimer" was filed in respect of them.

Patents were also granted to Aingworth (1844), Elliott (1844), Chatwin and Seymour (1845), Elliott (1845), Aston (1846), and others, but it would take too much space to describe them, and it is the less necessary to do so as a very handy *résumé* is to be found in the series of "Abridgments" of specifications relating to "Dress Fastenings," published by the Commissioners of Patents.

The linen or sewn-through button was patented, in 1841, by Humphrey Jefferies, who will be remembered by many as a famous swimmer, fencer, skater, and athlete in general. His button was not "three-fold," and he used copper for the metal ring, whereas zinc is now thought good enough. Many patents have been taken out subsequently for this useful article.

An extended survey of the button patents brings out very clearly the remarkable amount of ingenuity which has been expended in the production of an every-day article. If any of our readers will take the trouble to examine carefully a sample of each of the different sorts of buttons which he bears about him, he cannot avoid being struck with the genius—we might almost say—which has been displayed in their manufacture. The dissection of an ordinary cloth-covered button or linen sewn-through button reveals a composite structure which is the result of many operations, and the wonder is that they can be sold at so low a price and yet leave to the manufacturer a margin of profit which has laid the foundation of the fortunes of many persons in our busy town. With the exception of the covered button and flexible shank there is hardly an invention of importance in connection with the trade which has not arisen in Birmingham. The restricted area in which the trade grew up is very evident from the mode in which the specifications are drawn. As a rule every patentee seems to be perfectly well acquainted with the labours of his predecessors, and he refers to them in order to show more clearly the precise nature of his improvements. Such precautions are particularly necessary in the case of patents for small articles such as buttons, which have in many cases precisely the same general appearance, the novelty being only apparent after a very minute examination. The specifications, though in many cases very long, are models of clearness, and are generally illustrated by well-executed drawings which leave nothing to be desired. From the remarkable similarity in the mode of treatment, we should judge that a large number of them were drawn up by the same competent hand. Probably for the reasons given above, the number of old contrivances which have been repatented is not large, at least not so large as might have been expected from the nature of the subject. This remark does not, however, apply with so much force to the inventions of later years when we find many patents granted to "outsiders," who are obviously unfamiliar with the trade. The task of dealing with these must be left to posterity, as we do not profess to bring the history of the subject down to the present time.

## CHAPTER XIV.

### THE NAIL TRADE.

Cast Nails—Machine—Hammer Nails—Pressed or Cut Nails—Rolled-sheet Nails—Dyer's Nail Machine — Rolled-bar Nails — Metal-covered Nails—Church's and Prosser's Patents—Wire Nails, or "Pointes de Paris."

Although Birmingham has been the centre of the nail-making district of England for a couple of centuries, the inventive ingenuity of the town has done comparatively little towards devising machinery for replacing hand labour: one reason for this was perhaps the excessive cheapness of the article, which would naturally deter inventors from embarking in a costly series of experiments to produce a doubtful result. In America, however, it was widely different, and the dearness of labour, coupled with the large demand for nails to be used in the erection of the timber buildings so common in the New World, caused men to turn their attention to the question at a very early period. It is not we think very generally known that no less than 23 patents for nail-making machines had been granted in America before the year 1800. The present nail-making machine is practically the product of what may be termed the accumulated ingenuity of that country. The part which Birmingham took in the introduction of the trade is only second in importance to the actual invention of the machinery. It was here that the first experiments were made, and it was here that the difficulties which always attend a new manufacture were overcome. Before proceeding to deal with this part of the subject it will, however, be convenient to describe some of the earlier nail patents which have originated in the neighbourhood.

As regards nails made by casting there is not very much to record. A patent was granted to Joseph Ashton, in 1769, for a method of casting iron coffin nails and tacks which were subsequently tinned, and it would appear from his specification that cast-iron nails were then a novelty. He describes his process as follows :--"Melt pig iron, commonly called cast metal, in crucibles, in an air furnace, and run the metal when so melted into moulds filled with sand or loam, wherein the impressions of coffin nails and tacks have been first taken off from leaden or other soft metal patterns, and then neal the same nails and tacks in a fire made of coke from coal by a gradual heat for the space of twelve hours or thereabouts, whereby the same will become tough and malleable, and fit for use." The nails, after having been pickled and well scoured, were tinned by heating them over a fire in a vessel containing a portion of tin and a little sal ammoniac. In 1771 Ashton took out a further patent for nails, which he describes in the very same words as those above quoted; but whilst his first patent was confined to tacks and coffin nails, the second extended to nails of the following kinds: "sheathing, rose, sharp, trunk, bullion, hob, scupper, clout, Flemish, and sparrow-bill." There is no claim for tinning these nails. Ashton probably carried on business at the Moat, as we showed in one of the articles relating to buttons. Mr. R. F. Martineau, in his article on the cut-nail trade in "Birmingham and the Midland Hardware District," states that the "sparrow-bills" were so-named in consequence of the supposed resemblance of a certain kind of *cast* nail to a bird's beak, but it is evident from the above list that the first "sparrow-bills" were wrought. The next patent for casting nails is that granted to Richard Prosser in 1835, and differs from other processes in that plates are cast, having a section similar to that of a row of nails connected together head to point. These plates are next converted into malleable cast iron in the well-known manner, and then slit into rods, which are finished by being passed through suitably indented rollers, and finally divided by cutters into separate nails. Although the patent which was granted to John Holmes in 1838 does not strictly belong to Birmingham, we mention it here on account of the inventor's subsequent connection with the town. It relates to a method of preparing moulds for casting nails, amongst other things, and consists of a modification of the well-known "plate-moulding" process. The pattern for the inside of the head is on one plate, whilst that for the outside of the head and for the shank is on another plate. In the centre of each nail head pattern there is a hole, and when the box has been rammed up, the plate carrying the shanks is thrust through into the sand. Both plates are then removed and the two boxes brought together, when the mould is ready for pouring. In 1841, John Onions, of Darlaston (and subsequently of Bermondsey), patented a method of moulding by forcing a row of nails into a rammed box, by means of a particular arrangement of mechanism. The moulds for the heads are in a separate box. There is also a claim for nails of a triangular section, but the idea had been patented long before by another inventor. Mr. Samuel Kenrick's patent, granted in 1847, has already been referred to, under the head of buttons, as containing a mass of information on the subject of "odd-side moulding." We mention it again here on account of the description of the mode of preparing moulds for casting nails. We have been obliged to depart from the chronological arrangement usually observed in order to bring all the patents for cast nails together.

The earliest patent for the application of machinery to the manufacture of nails which was granted in this neighbourhood is dated 1790, and is in the name of William Finch, of Wombourne, near Wolverhampton. It consisted in little more than the use of hammers worked by power instead of by hand, and in the introduction of a system of division of labour, according to which a nail went through three hands before being completed. The patentee claims that by his method "the limbs of those employed in the manufactory will be preserved to the end of life, whereas, in the old method of working men are frequently lamed in seven or ten years, and obliged to be maintained by the parish." There is reason to suppose that the process was actually carried out.

It is stated in Rees's "Cyclopedia" that a very extensive trade was established, about the year 1804, in Birmingham and Sheffield, of brads cut in a fly press. This is the earliest mention we have found of the method of cutting brads in a triangular shape by placing the cutter at an angle to the strip, which is turned over at each stroke of the press, so so that every nail will be of the same form. The sideway motion of the strip, which is necessary when brads with heads are required, is also alluded to. Both of these methods have been in uninterrupted use from that day to this, and we can only regret that the name of the inventor has not been preserved. In 1808 a patent was granted to Joseph Willmore, silversmith, and John Tonks, plater, for a method of making nails by machinery. It was not a self-acting machine, such as we shall describe by-and-by, but rather a series of distinct operations, the first of which was the cutting of the nail rod into short lengths. The end on which the head was to be formed was cut square, whilst the point was cut bevelled on one side only. The next operation was the raising of the head in suitably-shaped dies, and this was followed by the pointing, which was also effected in dies. The machine employed throughout was the fly press, which was fitted with simple mechanism for closing the bottom dies as the heading punch descended, and opening them and releasing the nail on the rebound. The name of Tonks is, we need hardly say, very well known in the town, and the members of the Willmore family attained in former days some celebrity as engravers.

The idea of making nails by rolling—briefly referred to above—has exercised a great fascination on inventors, and has been patented over and over again in various forms. It was first started by Clifford, of Bristol, in 1790, and we find it revived in 1811 by William Finch, who is described as of "Birmingham, ironmaster." We assume this to be the same William Finch, of Wombourne, previously referred to. Clifford's plan was to pass nail rods through rolls each of which bore a series of recesses representing half of the nail, supposed to be divided longitudinally. The rod came out as a string of nails hanging together head and point, which merely required to be separated and finished in a die. Finch's specification is not quite clear, but he appears to use strips instead of rods, thereby forming what he calls "waved iron," from which nails are produced by "the use of tools commonly called the punch bed and bolster."

We have now arrived at the period which marks the commencement of the present cut-nail trade of the town. In the year 1810 Mr. J. C. Dyer, an American merchant, residing in London, obtained a patent on behalf of a company in Boston, which included several forms of nail machine, the invention of three Americans named Odiorne, Perkins, and Reed. Great difficulties were experienced in getting the machine into use, and a working model operating upon lead was taken about the country; but the Dudley and other nailmakers could not be induced to change their system. "Shortly after," says Mr. Dyer, in a paper read before the Manchester Philosophical Society, in 1861, "I succeeded in forming a company in London for establishing a patent nail manufactory on a large scale, to which company I transferred the patents, and undertook to superintend the building and starting of the machines for a period of six months; after which the concern was left wholly in charge of the company, the principal party in which was then an eminent London banker, who supplied the capital, and, as head of the concern, selected the parties to be charged with the management of the business. Besides a small sum received in money, I was to receive in compensation for the patent rights a certain share of the profits to arise from their exercise; but from the lack of mechanical knowledge and business talents that appeared in the management, as also from differences about the capital in the concern, I was induced to consent to an outright sale to the company of my contingent share of its profits at a price which did not exceed what my share ought to have produced *per annum*, if the affairs had been conducted with judgment and prudence." The company's manufactory, which was known as the Britannia Works, was situated in Blews Street, and amongst the persons connected with the business was Mr. Thomas Morton Jones, an American, who subsequently resided in the town for many years. When Mr. Dyer went out of the affair, he left a young American engineer, named Sewell Barker, who superintended the mechanical part of the business. He was an extremely ingenious man, and when the Odiorne machine failed on account of the numerous breakages which occurred, Barker took it in hand—remodelled it, and created what is, to all intents and purposes, the machine now in use. Mr. Thomas Pearson, who subsequently built a manufactory in Heneage Street, was also connected with the Blews Street Works.

The mechanical difficulties overcome, the next thing was to induce the public to buy the articles, and this proved very difficult. The London ironmongers refused to have anything whatever to do with the machine-made nails, but the authorities at the Britannia Works at last got hold of a Mr. Nichols, a retail ironmonger, in Bartholomew Close, and put him in business at No. 27, Old Bailey, which thenceforth became the London emporium for the new nails. At that time the Custom House was in process of rebuilding, after the fire of 1814, and as Mr. Nichols happened to belong to the same religious sect as Mr. Henry Peto, the contractor, he was able to secure a very large order. The nails were also very extensively used at Brighton. It is difficult to fix the date when they came into general use, but there is indisputable evidence that the nailers had begun to feel seriously the effects of the new manufacture by July, 1817, at which date a petition was presented to the House of Commons by the nailers of St. Ninians and Falkirk, setting forth that "so many machines through different parts of the kingdom are erected for making nails, by which nails are made to the ruin of the trade and devastation of the country; and some of which will make more in one day than a hundred men will make; and which machines are carried on by a few hands and pay no duty to government, particularly the machine for making nails called the Britannia, of Birmingham, and several others of the same description in different parts of the kingdom."

Mr. Dyer took out other patents for improvements in his machines in 1812 and 1814 respectively. The specifications include several distinct machines, which are described at very great length. They are exceedingly ingenious, and are well worthy of the attention of the mechanician, but it is unnecessary to describe them here, especially as they are not, properly speaking, Birmingham inventions. Mr. Dyer, who died in Manchester about fifteen years ago, was the introducer of several American inventions into this country. We may mention, amongst others, Amos Whittemore's card-making machines, the fisst of which was manufactured in Birmingham.

In the patent granted to Messrs. Daniel Ledsam and William Jones, in 1827, the cutters are mounted on a swinging frame, and the position of the strips is varied by the action of a sliding bar. Dr. Church's patent of 1827, taken out in the name of his correspondent Thomas Tyndall, includes a method of rolling rods with a series of depressions, which are cut up into lengths as they leave the rollers by the action of a descending cutter. The nail blanks are received into recesses in a rotating die which presents them in succession to a heading punch. In this machine we notice a peculiarly formed star wheel, which Church was very fond of, as he uses it in some of his other patents, for giving a "dwell" to the rotating die whilst the heading punch is doing its work. This machine in its main features resembles that for which the same inventor had a patent nine years previously. In 1831 Mr. Richard Prosser patented a coffin or bullion nail of iron, the head being covered with a thin layer of brass or other metal. A very neat and strong article was produced in this way, but we are not aware whether it ever came into general use, although an elaborate machine for making these nails was patented by Dr. Church in the following year. Messrs. Ledsam and Jones appear in 1831 as patentees of a machine in which the wire or rod is drawn into the apparatus intermittently, cut off to the right length, held in dies with a small portion protruding, which is then operated upon by a punch and caused to spread laterally to form the head of the nail. The ingenious Dr. Church, who we omitted to say was also an American, follows in 1832 with another modification of his method of forming nails by rolling. In another machine, which is described in the specification, the points of the nails are formed by cutting off pieces of metal obliquely, in such a manner that the portions removed form shoe-bills. Mr. Richard Prosser's machine, patented in 1835, is arranged to make nails from rods heated in a small furnace (kept hot by a blast) situated in front of the machine. A length of the heated rod is drawn into the machine, the tapering of the shank and pointing being effected by two pairs of hammers or plungers, which act simultaneously at opposite sides of the rod. It is next cut off, seized by a carrier, which places it in the heading die, and subsequently discharged as a finished nail. The combinations of mechanism contained in this machine exhibit a considerable amount of ingenuity. A heading apparatus, consisting of a wheel with radial orifices revolving in contact with another wheel carrying a corresponding series of heading dies, is also described. The patent likewise includes a rolling process very similar to those already referred to. A rotating heading die is also the feature of Robert Griffiths's patent of 1835. In the following year William Southwood Stocker took out a patent in which the hammers for pointing the nail work in pairs, very similar to Prosser's machine just described, so that no fixed anvil is necessary. Alexander Stocker and Henry Downing's patent of 1836 relates to a particular mode of forming the dies for nail, screw-blank, and rivet machines of cast iron. In W. S. Stocker's patent of 1840 we find a method of drawing down the stem of the nail by the use of grooved rollers, the bearings of which are caused to advance towards each other as the rod is drawn through, an idea which found favour with several subsequent patentees. Another modification relates to cut nail machines, and consists of a method of turning the strip at each stroke of the cutters.

The specifications of many of these nail patents are very elaborate, but De Meckenheim, 1847, certainly bears away the palm. The inventor was an Austrian who was settled in the town for a short time, and his patent contains machines for making nails by every known process. He has a method of rolling nails, an ingenious device for reversing the strips in cut-nail machines, and apparatus for making "Pointes de Paris," or wire-nails, besides a host of contrivances for pointing and heading, and for making screw blanks and bolts of all sorts. In 1849, William Goose patented a method of turning the blanks as they are cut, so that the compressing dies shall operate on the sides of the nails instead of their edges, and thus avoid the liability of opening the grain of the metal. John Manly, junior, in 1850, had a patent for a method of ornamenting the heads of nails by stamping, and also for covering them with various substances, such as thin metal, paper, cloth, flock, horn, silk, &c., in a similar manner to that employed when covering buttons. Messrs. Leonard Bower and Thomas Fortune's patent of 1850, relates to a method of tapering the shanks by pressure between dies, by which the work of the cutting and pointing dies is facilitated. We conclude the series with the patent of Messrs. John Hinks and Eugene Nicolle, granted in 1852, for a machine for making wire nails, in which the material is first subjected to the action of rollers, then separated by cutters, each nail being finished separately in dies.

## CHAPTER XV.

### THE PIN TRADE.

Early Pin Makers—Ryland—Phipson—Solid-head Pins—Wright's Machine—Heading and Pointing Machines — Japanning Pins — Nursery Pins — Machine for Papering Pins.

It appears to be a well ascertained fact that the pin trade of the town dates from the middle of last century, when it was commenced by the Ryland family. In the "Directory" for 1770 we find amongst the "Miscellaneous Tradesmen" the name of "Samuel Ryland, pin maker, No. 16, New Street." The business was transferred in 1785 to his nephew, Thomas Phipson, who carried it on for some years at the same place, as appears from Bisset's "Directory," published in 1800. We have nothing to record as to the application of local talent to the improvement of this particular trade until a comparatively recent period, and the manufacture of a pin has served on so many occasions as an illustration of the advantages of the "division of labour," that we need not describe the old process here. It has probably also done duty as the text of a homily on the virtue of patience under the minor provocations of life when—as very often happened before the days of the "solid-headed" article—the head gave way and the shank ran into the fingers. Such accidents rarely, of course, happened with the very superior pins for which Birmingham has always been famous.

The first mention of the solid-headed pin is, we believe, to be found in a patent granted in 1817 to Seth Hunt, an American then residing in London. The patent was for a self-acting machine for making solid-headed pins, but it was not Hunt's own invention, having been communicated to him by a foreigner residing abroad, in all probability a fellow-countryman. Nothing appears to be known as to whether Hunt's machinery ever came into use ; and no other patent was granted for pins until 1824, when Lemuel Wellman Wright, a native of Plainfield, State of New Hampshire, where he was born in 1789, obtained protection for a machine which eventually became of great importance. As will be readily understood, these machines are of very complicated construction, not altogether dissimilar in arrangement to some of the nail-making machines described in our last article. Wright's machine has been very frequently described, and a philosophical analysis of it may be found in Babbage's "Economy of Manufactures," together with a minute account of the old method of making pins. It was first set to work at his factory in Wellclose Square, where it formed one of the sights of London, according to the writer of the volume relating to "Manufactures in Metal," in Lardner's "Cabinet Cyclopedia." The manufacture did not, however, prove a commercial success ; and the patent was acquired by Messrs. D. F. Taylor and Co., of Light Pool Mills, in the Vale of Rodborough, near Stroud (now of this town); but it was not until 1833—nine years after the grant of the patent—that the solid-headed pin was introduced into the market. This statement is based upon a paragraph in the *Mechanics' Magazine* for November 23, of that year. It does not seem to have created any very surprising sensation, and a long period elapsed before the old-fashioned article was driven out of the market. Wright's patent was extended by the Privy Council for five years, so that it finally expired in 1843. A machine was produced before the Lords of the Council and worked in court, and evidence was given that no less than £20,000. had been expended before the manufacture became profitable.

Gloucester has always been the seat of an important trade, and we find it stated in the *Mechanics' Magazine* for April 7, 1838, "that in some years pins to the value of £20,000., or at least 300,000,000 in number, have been sent to the Metropolis from Gloucester alone, during the twelvemonth." On the 26th of October, 1839, the same journal notices the establishment of a large pin factory at Latchford, near Warrington, by Messrs. John Edlesten and Son, which, when in full work, would give employment to 1,000 persons. The average weekly production at that time was said to be between fifteen and sixteen millions.

Although we are not able to fix the precise date when the manufacture of solid-headed pins was actually commenced in Birmingham, the first patent relating to the subject was granted to Daniel Ledsam and William Jones, in the year 1831. Two separate machines are described, one for heading, and the other for pointing. When the first machine has done its work the headed pins are fed from a hopper on to the periphery of a wheel, having a series of grooves across its face and of course parallel to the axis. The heads of the pins lie in these grooves, and the revolution of the wheel brings them in contact with a curved bar, which, like the periphery of the wheel, is covered with wash-leather. A rotary motion is thus given to the pins whilst they are being sharpened by contact with a rapidly revolving file. Some further improvements in the machine formed the subject of a subsequent patent by the same parties in 1833. The main feature was the addition of mechanism by which a more rapid rotation was given to the pins during the operation of pointing. Although "pins" are mentioned in the title of the patent granted to William Southwood Stocker in 1840, it is evident that the machinery there described is not applicable to the manufacture of the "pins" now being dealt with, but rather to the making of the larger articles known by that name, and more allied to rivets. The patent of Mr. George Goodman, of Henley-in-Arden and Birmingham, dated 1840, relates to mourning and dress pins which the inventor makes from "iron, common steel, or cast steel wire." The operations are generally similar to those followed when forming needles, but when iron wire is used the pins when partially completed are converted into steel by cementation. They are headed in the old way, and may be either "blued" or blacked. The dress pins referred to are headed either with silver or plated metal. Mr. W. W. Jenkins of Greet, near Yardley, patented a machine in 1841, in which a number of blanks ready pointed are fed from a hopper and caused to pass one by one into a groove in a metal plate. The blank is pushed by a plunger into the jaws of the holder, which then travels laterally till it comes opposite the heading dies. Two operations are necessary to complete the formation of the head. The patent also includes a machine for sticking pins into paper, which will be noticed further on.

De Meckenheim's patent (1847), which was referred to in the article on nail-making, contains also a machine for making pins. The wire is fed in by rollers which have a continuous motion, but portions of their surfaces are cut away in order that they may advance the wire only during part of their revolution. When the head has been formed it is subjected to the action of the pointing tools, which consist of two discs having curved files attached to them. By the rotation of the discs the curved files take hold of the partially-formed pin and compress and file the point. The files are not concentric with the axis of the discs, but have an eccentricity, in order that they may compress and form the point gradually. A patent was granted to Mary Jenkins, of Aston, in 1847, for japanning pins, amongst other articles, of various colours, black excepted.

Mr. Charles Rowley's nursery pin, patented in 1849, has been "a real blessing to mothers," and it possesses, as the inventor says, "extreme simplicity and economy, combining security with safety." Without in any way detracting from the ingenuity of the contrivance, its absolute novelty is somewhat open to question, for a very similar method of protecting the points of the brooches worn by the Roman ladies was in use nearly two thousand years ago, as may be seen by the specimens of fibulæ preserved in the British Museum. There is, however, just this difference, that the safety pin is made out of a single piece of wire. The same specification contains various methods of protecting the points of dress pins by a small sheath kept in its place by means of an elastic connection formed either of indiarubber or by a hardened bent wire. In the following year Mr. Rowley took out another patent embodying several fresh forms of point-protectors, which are characterised by the peculiar ingenuity manifested in nearly all the inventions of this patentee.

The machine patented by Mr. Samuel Lusty, in 1852, marks a considerable advance towards simplicity in construction. The wire is fed in by a lever actuated by a cam and seized between two dies carried in a frame which traverses laterally to sever the wire and bring it in front of the heading punch. The blank is then released and allowed to fall into an inclined channelled trough provided with stops to prevent the blanks falling too fast and to ensure that the heads shall always be uppermost. The blanks hang by their heads in the groove in the trough and are led to the pointing apparatus, which consists of a rotating file, the blank being caused to revolve whilst it is undergoing the operation by means of a strip of wood. In this manner an even point is given to the blank, and when it has traversed the length of the cutter it is discharged into the receptacle a finished pin.

Who first proposed to stick pins on to paper for sale is a question which must, we fear, remain in obscurity. The notion was excellent, and when pins were dear the cost of the operation could not have borne a very serious proportion to the value of the articles themselves. But as the manufacture increased means were sought to dispense with the expensive hand-labour employed, and machines were introduced. A patent was granted in 1839 to Miles Berry, a London agent, as a communication from a foreigner residing abroad for a contrivance of the sort; but a much better machine was patented by Mr. Jenkins, of Greet, in 1841. The pins to be stuck are placed in a hopper with the heads all one way. The hopper is caused to pass over a plate having a number of grooves corresponding to the number of pins in the row, by which operation a pin is deposited in each groove. In front of this plate there is a ribbed bar, also grooved transversely to correspond to the plate. The paper is hung over this bar and a second bar pressed down on it, thereby crimping the paper between them. Whilst the paper is so held the row of pins on the grooved plate is pushed through the folds in the paper, and the operation repeated on another part of the sheet.

## CHAPTER XVI.

### THE NEEDLE TRADE.

WEISENTHAL'S PATENT — SHEWARD'S DRILLED EYE NEEDLES—CAST NEEDLES AND BODKINS—GILT-EYED NEEDLES — NEEDLE-POINTING MACHINERY — LEDSAM AND JONES'S PATENT — MORRALL'S PATENT FOR BURNISHED-EYE NEEDLES.

STRICTLY speaking this has never been a Birmingham trade at all, nor are the patents for improvements in needle-making which belong to this town very numerous. In fact the manufacture never owed much to patents, and there are comparatively few entries relating to the subject, except of late years. The first is that granted in 1755 to C. F. Weisenthal, a merchant carrying on business in London, for a double pointed needle, the eye being in the centre, thus avoiding the necessity of turning the needle in doing embroidery work. This could hardly have been a new idea even then, and it has been patented about half a dozen times since.

Weisenthal's patent was, however, only for a new form of needle, without any reference to the mode of manufacture. The earliest grant of the kind is dated 1775, and is in the name of William Sheward, of Redditch, "worsted needle manufacturer," for needles with eyes upon a new and particular construction." The needles being eyed and guttered the inventor brings them in contact with "a small steel tool, drill, or rimer," carried in a lathe. The tool is kept revolving, and "the eyes of the needles are then put upon or to the point of the said drill or rimer, in such manner and form, and continue there so long till the said tool, drill, or rimer hath worked in and about the eye of the needle, so as to take off and smooth all the edges from the eyes of such needles." Here we have the first attempt at making "drilled-eyed" needles, an expression which does not, however, denote that the eyes are actually drilled out of the solid. It is stated that a Studley needle-maker introduced this method in 1793, but, as might have been expected, it failed, on account of the expense. Sheward seems to have subsequently removed to Birmingham, for in 1789 he had a patent, in which he is described as of that town, for a new and improved method of finishing the eye of a needle "in a way superior to any yet invented." His invention consisted in arranging upon a revolving shaft a number of steel discs, alternately roughed and smooth, the former being larger in diameter than the latter. The needles being applied to the roughed discs were rendered free from the "square edges that punches always leave in the eye of a needle, and the small wheels will burnish or polish the cheeks and full parts about the heads and turn them off complete." The only information we have been able to gather respecting Sheward is that he at one time occupied a farm at Peoley, near Redditch, and that he was also the inventor of a particular form of horizontal wind-mill, the sails of which were inside a hollow drum. This was an idea which found particular favour during the last century.

The next patent to be noticed is that granted in 1795 to William Bell, of Walsall—who, we have reasons for thinking, was an importation from Birmingham—for making "all sorts of needles, bodkins, knitting pins, fish-hooks, netting needles, mesh pins, and sail needles." The method consisted in casting the articles in sand moulds, the metal used being partly refined by melting with charcoal dust and lime or common salt, the mixture being frequently stirred with an iron rod. The inventor states that articles cast in this manner are capable of being tempered and hardened or softened in the usual way. However suitable the process may have been to the production of bodkins and such like articles, it would be totally inapplicable in the case of the finer kinds of needles.

In 1812, John Scambler, a needle-maker of this town, introduced the "Patent Golden Needles," the chief peculiarity of which was that the eyes were gilt by being dipped in a solution of gold in aqua regia, a process which at best would give but a very unsatisfactory result. The needle had also a long point, the taper part commencing at about two-thirds the distance from the extremity. The eyes were square or round instead of oval, and the needles were hardened by immersing in oil and water.

It is often asserted that the needle-pointing machine is of German invention, but that is an entire mistake. The invention is English, and the error may have arisen from the fact that needle-pointing machinery was first used in a German manufactory, the unreasoning opposition of the workmen preventing its introduction here for many years. Pointing by hand is however now almost entirely obsolete at Redditch, foreign competition having compelled manufacturers to adopt the machine. The needle-pointing machine is much older than is generally supposed, the original patent having been granted in 1833—nearly half a century ago—to Daniel Ledsam and William Jones, of this town. We have had occasion to mention these inventors before in connection with the manufacture of nails and pins. The machine in question not only grinds the points of needles, but it cuts off the wire, the lengths being sufficient for two needles. The end of the wire, which is coiled on a drum, is passed through a set of straightening pegs, and is then seized by a pair of pliers, which draw off a determinate length of wire from the coil. The amount thus drawn off at each pull of the pliers is governed by a crank and slotted arm, the bearings of which may be altered according to the "throw" required. The wire is then severed by means of a sliding cutter, and the part cut off held in a tube. Adjacent to the end of this tube there is a wheel with grooves in the periphery, each capable of holding a single needle. When a fresh length of wire is cut it pushes that previously severed, and which has been described as remaining in the guide tube, on to the groove in the wheel, the face of which is not broad enough to take the whole length of the needle, so that the ends overhang. A bar, bent to the curve of the wheel, keeps the needles from falling out of the grooves as the wheel revolves, and also serves to impart a rotatory movement to each needle by the friction between the wheel and bar, the friction being increased by covering both with wash-leather. In this position they are presented by the continued revolution of the wheel to the grindstone, which is hollowed out to correspond to the curvature of the former. A projecting guard-plate gently depresses the needles to bring the points in more certain contact with the stone. The grindstones, which are driven independently and have special modes of adjustment, are two in number, one on either side of the wheel, so that both ends of the wire are sharpened. When the operation is complete the eyes are punched out, and the needles are finished in the usual way. We are not aware whether Ledsam and Jones's machine was ever used, but its main features have been reproduced in that now generally employed, which is, however, not entirely self-acting, the rotation of the wires being imparted by the fingers of the grinder.

A most important invention was patented in 1839, by Abel Morrall, a Studley needle-maker, for burnishing the eyes of needles by threading them upon a roughened steel wire stretched in a frame and caused to revolve, or to move backwards and forwards. The needles are thus made to vibrate upon the wire in every direction, and the eyes effectually cleared from all roughness. This very valuable patent was shortly afterwards purchased by Messrs. Bartleet, of Redditch, and the use of "string or cord" which the inventor thought might also serve as well as wire was disclaimed by them in 1841. An attempt was made by a league of twelve rival manufacturers to set the patent aside by a *scire facias,* but the proceedings were unsuccessful, and the patent was held to be valid. The effect of the introduction by Messrs. Bartleet of the oval-eyed needles perfected by this machine has doubtless been to cause the eyes of needles generally to be made larger than they formerly were, to the great convenience of the majority of persons who use them. Up to that time there were no means of making the eyes perfectly smooth, except in the case of round-eyed needles, and even they were benefitted by the use of the burnishing machine. But it is still a question if any needle is so good in use as one with a perfect round eye, carefully drilled, and then burnished by the best method known in the trade.

In 1841, Luke Hebert, a consulting engineer and patent agent, then residing in Birmingham, took out a patent for a machine which was a palpable imitation of Morrall's. Hebert proposed to string the needles on a roughed wire, as in the former machine, but the needles were to be firmly held in a clamp whilst the wire was drawn backwards and forwards through the eyes. The plan failed in practice, as might have been foretold, for it did not produce the right kind of action upon the eyes.

## CHAPTER XVII.

### THE MANUFACTURE OF METAL TUBES.

SHEET LEAD PIPES — CAST LEAD PIPES—GROOVED ROLLS—CIRCULAR SHEARS—WYATT'S COPPER PIPES —SQUIRTED TUBES—CASED POTTERY TUBES—MANDRIL DRAWN TUBES—ROLLED TUBES.

With one exception, which we shall refer to presently, the various inventions connected with the metal tube trade are of purely indigenous growth, and not a single patent of importance has been granted outside the town and district, at all events up to the middle of the present century. The inventor of the drawbench, which in all probability originated in Germany, is not known, and the date of its application to the manufacture of tubes is equally uncertain. Drawn tubes were, however, in common use in 1782, as we learn from a patent granted to one Joshua Lover Martin in that year for a method of drawing plated metal tubes for telescopes and other philosophical apparatus. This was probably the earliest use to which they were applied, although there must have been some demand for copper piping for distillers' worms from a very early period.

Lead pipes were largely used by the Romans for conveying water to their houses, and a learned paper was read before the Paris Academy of Sciences about fifteen years ago on the various fragments of ancient lead pipe which had been from time to time discovered in that city. In the middle ages lead pipes were made by bending a sheet so that the edges touched, and then pouring molten lead on the outside of the pipe to form the seam, the position of which is marked by a projecting rib. The seams may, perhaps, have been soldered in some cases, but the method we have described was certainly followed, for we have in our possession a piece of pipe taken from the ruins of Furness Abbey, the joint of which is of the same composition as the other part of the pipe, and does not contain a trace of tin. A great improvement in the manufacture of lead pipe was patented in 1790, by the celebrated John Wilkinson, of Bradley, the father of the iron trade, as he was wont to call himself, and who contributed in no small degree to the success of Watt's steam engine by his skill in casting and boring cylinders. The metal is first cast into the form of a short tubular ingot, which being placed upon a mandrel is reduced in diameter either by being passed through grooved rolls or by the draw-bench. There is, however, a still earlier instance of the use of grooved rolls in connection with the manufacture of lead pipe, and it is contained in a patent granted in 1741 to James Creed, of East Greenwich. The only use of the rolls is to turn up the plates to the required form, previous to soldering the seams. A particular form of circular shears, an implement now in almost universal use for a variety of purposes, is also included in Creed's patent. Wilkinson's method, which is still, we believe, employed, is of considerable interest as containing the germs of two most important inventions which subsequently arose. Grooved rolls were shortly afterwards applied to the welding of gun barrels and then to gas pipes, whilst the idea of producing seamless brass tubes by drawing a tubular ingot was introduced with great success by Mr. Charles Green, in 1838. These remarks must not be understood to detract from the merits of the various inventors who have worked in these directions, as the difficulties of applying an old method to a new material are often so great as to give an undoubted claim to the introduction of "a new manufacture within the realm," as the statute of monopolies has it. These developments will be referred to in their proper place later on.

We may here record the fact that a considerable trade appears to have been carried on in copper pipes by one Charles Wyatt, merchant, who had a patent in 1791. There was no improvement in the actual mode of forming the pipes, which were to be manufactured by the ordinary methods then in use, the claim being for the application of copper coated with tin, or tin combined with lead, to the manufacture of pipes, troughs, spouts, cistern heads, &c. In conjunction with one Collins, who has been mentioned more than once in these articles, Wyatt had previously obtained a patent, in 1790, for methods of coating copper and iron with tin and compounds of tin. In a circular issued by the inventor he says that his pipes "have greatly the advantage, it is presumed, of other metallic pipes, from the properties which they possess in common with the sheets. They are stronger, lighter, and very considerably cheaper than lead; unquestionably sweet; less expensive, and infinitely more durable than wood. They form a cheap substitute for pump-trees, and may be had of all dimensions and of sufficient strength to resist any given pressure of water. Prices of under-ground pipes delivered in lengths of six feet each or less, warranted to support a column of water 100 feet perpendicular (about 43lb. to the square inch): ½in. diameter, 5d. per foot; ¾in., 5½d.; 1in., 7½d.; 1¼in., 9½d., 1½in., 1s.; 1¾in., 1s. 2d.; 2in., 1s. 4d.; 2½in., 1s. 6d.; 3in., 1s. 8d.; 3½in., 2s. 2d.; 4in., 2s. 6d. per foot." According to an advertisement inserted in the *Monthly*

*Review* for December, 1793, Wyatt also made "leaded iron pipes for rain water, and also for conveying water underground, at prices varying from 5d. per foot for pipes 1in. diameter, to 1s. 9d. per foot for those 4in. diameter. He also made them with flanges, "of sufficient strength for supplying towns with water, at about two-thirds the price of cast-iron pipes." We have a suspicion, which we should be very glad to convert into a certainty, that this Charles Wyatt was a son of John Wyatt, the associate of Lewis Paul in the cotton-spinning scheme which we described in the second article of this series. We know that he had a son named Charles, and another son (William Henry) was settled in London, where he was the publisher of the *Repertory of Arts*. What seems to lend force to the supposition, is that Charles Wyatt's agent in London was also a John Wyatt, who we take to be his brother.

Allusion was made at the outset to a certain process for making metal tubes which was of foreign origin. We referred to those made by pressing the material through an annular orifice, which has long been practised in the town. Up to this time Thomas Burr, plumber, of Shrewsbury, who took out a patent for making pipe in this manner in the year 1820, has always been credited with the invention, but it was really due to Bramah, although we have never seen it alluded to in any of the memoirs of that distinguished mechanician. Mr. Smiles has overlooked it, although he mentions Bramah's patent of 1797, which contains the invention in question. There is nothing whatever in the title of the patent to lead one to suppose that it contains anything on the subject, for it relates to raising beer from cellars on a principle which approaches very closely to that of Sir William Armstrong's "accumulator." It must not, however, be supposed that the specification merely contains a suggestion, as it is obvious from the wording of the document that Bramah had actually tried the experiment. He may or may not have taken the idea from the macaroni makers; and there is just this difference between his process and that now followed, that he keeps the metal in a molten condition and chills it as it issues from the die. His object appears to have been simply to make piping for conveying beer from the cellars of taverns to the bar; but it is curious that the invention should have been allowed to slumber for so many years, especially when a large demand arose shortly afterwards for cheap and safe tubing for the purpose of conveying gas for illumination. It has always been said that old musket barrels were used during the early experiments at Soho, and this is no doubt true, but it appears from a letter in our "Local Notes and Queries," of the 5th inst., that thick tin-plate was also employed, the joints being made with duck coated with white lead. A piece of the original tubing was to be seen at the Gas Office some time ago, but unfortunately it has been lost.

Although not strictly a metal tube, we may here mention William Bell's patent of 1808, for a method of enclosing tubes of porcelain or pottery in iron tubes for the purpose of protecting them, and to prevent breaking or bursting. At that time there were great difficulties in the way of obtaining tubes for water supply. Wood had been used, but it was liable to decay and to generate insects, although it had a tolerably long run in popular favour. There were several patents for boring wood pipes, one of which was granted to Joseph Chirm, a pump-maker of the town, in 1801. Cast-iron pipes were only to be had of small diameter and in short lengths, and there seems to have been an unfounded prejudice against the material on account of its action upon the water. Bell's patent may be regarded as a sort of precursor of the enamelled iron pipes, which were not introduced until many years afterwards.

The method of drawing a tube of brass or other metal over a tube or rod was first introduced by Sir Edward Thomason, in 1803, as is stated in his "Memoirs." Struck by the fact that it required as much force to slide a tube from a mandrel as was necessary to draw it on, he commenced to make experiments, and found that two metals such as copper and iron, or brass and iron, could be made to adhere firmly by pressure. He applied his invention to the manufacture of stair rods, picture-rods, &c., but he did not take out any patent for it. In 1812 Thomas Attwood and Benjamin Cook patented the very same thing, but Thomason had no difficulty in proving his prior claim. A further development in the same direction was patented by J. W. Phipson, in 1819, consisting of a combined lead and copper pipe, one being drawn over the other, principally for the purpose of conveying gas, which was still regarded with suspicion and not fit to be admitted into houses except under the most stringent conditions. The internal portion was of lead, as being proof against the corrosive action of the gas, the copper being outside for the purpose of giving strength and presenting a more sightly appearance. Cased piping was also repatented in 1841, by Job Cutler, who describes a method of drawing and rolling combined iron and copper tubes for locomotive boilers. In the course of a specification, dated 1820, having reference to coating lead with tin, Thomas Dobbs describes a method of tinning lead pipes by casting a thin layer of lead either upon the inside or the outside, the surface having been previously prepared by rubbing molten tin on with a wisp of tow, a little rosin being used as a flux. Dobbs was the inventor of the celebrated "Albion metal," which was so frequently before the courts during the litigation connected with Betts's patent capsules about twenty years ago. In his first patent, granted in 1804, he is described as a "chemist, of King's Norton," but he appears to have subsequently removed to Smallbrook Street, where he carried on the business of a plater. He must not be confused with James Dobbs, long a favourite comedian at the Theatre, the inventor and patentee of a reaping machine.

The mode of turning up skelps for taper tubes, patented by Thomas Breidenbach in 1828, is curious, as being intermediate between rollers, and moulds extending the whole length of the skelp, and acting like a stamp or die. The apparatus consists of two moulds, each having a semicircular recess, so that when placed together a perfect circle is formed. The plate of metal is laid upon the lower mould and forced into the groove by a long mandrel, so that it assumes a trough-like form. The upper mould, which is kept in contact with the lower one by sliding in a V groove, is then moved over the imperfectly formed tube, the edges of which are thus bent down into a perfect circle. Several modifications of this plan of turning up skelps by the use of moulds were patented subsequently.

We have departed from the chronological arrangement usually followed in order to enable us to deal with the brass and iron tube patents consecutively. This we hope to do in the next article.

---

## CHAPTER XVIII.

### THE MANUFACTURE OF METAL TUBES (CONTINUED).

GREEN'S PATENTS — CALICO - PRINTING ROLLERS — VARIETIES OF TUBES—WELDING JOINTS OF TUBES—PUSHED AND DRAWN TUBES—REMOND AND PROSSER PATENTS—ROLLING TUBES—WARD'S DRAWN TAPER TUBES—FLUTED AND SPIRAL TUBES—BENDING TUBES HELICALLY.

THE successful production of solid or seamless brass or copper tubes is entitled to rank amongst the most important inventions connected with the trade. It was not an improved mode of producing a known article, but it was an entirely new thing. The invention is due to Mr. Charles Green, whose patent is dated 1838. It consisted in drawing a thick tubular ingot, the internal diameter of which was the same, or nearly the same, as that of the finished tube, until it had been reduced to the required thinness. The drawbench used was the same in principle as that formerly employed, but it received some alterations in point of detail which greatly increased its power. Instead of the ordinary solid dies, the inventor also proposed to use four rollers arranged at right angles, the periphery of each being hollowed out, so that when brought together a complete circle was formed. This was a very ingenious adaptation of the tube-welding rolls already in use, but the action was not a rolling action, since the grooved rolls were only rotated by the tube being drawn through them. This may seem to be a very small difference, but a patent was nevertheless granted to Thomas Smedley, of Holywell, in 1840, for a method of *rolling* tubes from thick tubular ingots, the only difference between his and Green's process being that the rolls were rotated by bevel gearing. In the same year Alexander Southwood Stocker, of this town, patented a method of rolling tubes scarcely distinguishable from that patented by Smedley, except that he used an alloy of copper and zinc of the proportions of "Muntz's metal," with the addition for certain purposes of a small percentage of tin. He also proposed to make seamless tubes of malleable iron in the same manner.

In 1841 Mr. Green obtained another patent for various improvements in his process, including the casting of a rim inside the ingot to form a stop for the mandrel, which is furnished with a corresponding shoulder. The patent also included a method of making boiler tubes thick at one end, so as to stand the increased heat to which they are exposed near the furnace. This was effected by casting the ingot smaller internally at one end and reducing the size of the corresponding portion of the mandrel. By another application of the method he produced shells for calico-printing rollers, which are merely thick copper tubes, having an internal projecting rib, which takes into a corresponding slot in a solid iron mandrel, and thus prevents the shell from turning round.

We digress here for a moment to point out that this method of fixing the shells of calico-printing rollers, known as the "slot and tab," was patented by Jonathan Ridgway, a plumber, of Manchester, as long ago as 1815. Many attempts have been made to supersede it, but we believe that none have been entirely successful. The intrinsic value of the copper rollers in an extensive print works is very large, and as the surface only is required, it was proposed by the late Richard Prosser, of this town, in 1853, to use a plain tube or shell of not more than the tenth of an inch in thickness. The shell was fixed by making the mandrel taper, an elastic ring being interposed between it and the shell; but when it was keyed up the pressure of the ring was found to throw the cylinder out of truth. This distortion was fatal to the success of the plan, it being essential that calico-printing rollers should be accurately cylindrical. Much money was spent in experiments, and Messrs. Hoyle and Co., of Manchester, were ready to take the matter up had it proved successful. The tube-makers of the town have paid considerable attention to the manufacture of printing rollers, and we find that the well-known Benjamin Cook was a patentee of a method of drawing copper tubes on iron mandrels for this purpose in the year 1817. This may be regarded as a further development of his — or rather Sir Edward Thomason's—method of making cased-iron rods, described in the previous article. Cook also took out a patent in 1829 for fixing the shell on an iron mandrel, made slightly oval, so that the shell would not slip round. In 1830 he patented a mode of forming the rib by cutting away the surrounding metal. In 1850 Mr. George Attwood patented a plan of making tubes out of old printing rollers by boring out the rib or slot from the interior, and after placing them on a steel mandrel drawing or rolling them down to the required size. In the early days of calico-printing the rollers were cast solid, rolled through grooved rolls, and finished by swaging. The interior was bored out, and the rib produced by inserting a slotted mandrel and passing through grooved rolls, the process being assisted by hammering the roller, so as to force the metal into the recess in the mandrel. This somewhat rude method was patented in 1825 by James Lockett, of Manchester. This digression will serve to show how extensive are the ramifications of the tube trade, which is not confined strictly to the production of long hollow cylinders intended for the conveyance of fluids or gases. We must return, however, to the manufacture of tubes properly so called.

In 1848, James Rose and William Haden Richardson, both of Darlaston, patented a machine for rolling brass and copper tubes from a short tubular ingot, the only new feature of which appears to be the use of grooved "supporting guide rollers" at the back and front of the grooved rolls used for extending the tubes. In the following year Job Cutler, of Birmingham, had a patent for making brass and similar tubes in the same way, but instead of employing a single pair of grooved rolls with a series of gradually diminishing holes, the patentee prefers to use a series of rolls, the axes of which are parallel, the grooves in each pair being smaller than those in that immediately preceding. By this arrangement the tube passes straight through the rolls and may be finished at a single heat. The mandrel is released either by hammering, or by a system of "cross rolling." The numerous patents which have been taken out for rolling tubes from a short and thick tubular ingot furnish an interesting instance of the prevalence of what may be described as "fashion" in invention. There are, it is true, some small points of difference, which we have not thought it necessary to refer to, but in their main features they are almost identical.

Reference was made in the previous article to a curious method of making lead pipe, which was practised in the Middle Ages, somewhat analogous to the process known in the foundry as "burning-on." A similar plan was patented in 1849 by James Banister, who proposed to unite the edges of a skelp turned up into the form of a tube by running molten metal along the seam, the tube being filled with sand, and a ridge of the same material being laid on each side to prevent the metal from spreading. The tube is to be made red-hot previous to the operation, and the molten metal, which is of course to be of the same composition as the tube itself, partially fuses the edges of the skelp and makes a perfect join. The solid ridge formed by the superfluous metal is afterwards removed by a saw, and the tube finished in the usual manner. Banister was also the patentee in 1851 of a particular construction of muffle for brazing tubes, so contrived as to diffuse the heat uniformly.

The process of making tubes which was patented in 1851 by Amédée François Rémond, an ingenious Frenchman who was living in the town about thirty years ago, was said to be applicable to making tubes of all kinds, but we notice it here because it was applied to the manufacture of the copper shells for calico printing rollers mentioned previously in connection with Richard Prosser's patent. The process consisted in gradually "dishing" a circular plate of metal by the action of a succession of pairs of dies, until it assumed a cup-shaped form with parallel sides. A mandrel having been inserted, the partially-formed tube was pushed, not drawn, through a series of dies gradually diminishing in diameter, or a series of grooved rollers might be employed. When the tube was sufficiently reduced, it was removed from the mandrel and the closed end cut off. The specification describes a very well-arranged machine for performing this operation, the construction of which will be understood if we describe it as a direct-acting steam draw bench. At one end of the bed a horizontal steam cylinder is placed, and the mandrel is connected with the piston rod in such a manner that the power is applied direct. The extent of the stroke may be varied according to the length of the tube on the mandrel. Considerable care was required to prevent the metal from puckering during the preliminary operation of dishing, and it was very liable to crack at the edge. Ingenious as the process might have been, it was, in fact, little more than a revival of a method patented in 1808 by Benjamin Cook, for the manufacture of gun-barrels, which Rémond likewise mentions amongst the objects capable of being made under his patent.

In 1851 Mr. Charles Green patented a method of making brass boiler tubes soft at the ends so as to admit of being easily fixed in the boilers, and at the same time possessing smoothness and stiffness, which latter quality is of importance as adding strength to the boiler.

Mr. George Frederick Muntz, junior, introduced in 1852 an entirely new process for making seamless tubes of the metal with which his name is associated, or of any similar metal which will roll hot. A tubular ingot is first cast according to the ordinary method, but of an oval instead of a circular section. This ingot is heated to redness and passed through rolls which flatten it, suitable precautions being taken to prevent the adherence of the internal surfaces. When the ingot has been sufficiently extended it is opened by a pointed mandrel, and the tube is finished in the usual manner.

The manufacture of taper tubes was first effected by the use of rolls with a groove of a gradually diminishing depth, and many patents were taken out for making gun barrels according to this method. In 1841 a patent was granted to Dr. Church and Jonathan Harlow, of the well-known firm of Peyton and Harlow, of Bordesley, for the manufacture of taper brass tubes for bedsteads on this principle, and Thomas Deakin appears as the patentee of a somewhat similar method in 1850. The problem of *drawing* taper tubes was, however, first solved by Mr. John Ward—then a workman in the employ of Messrs. Winfield—wh obtained a patent in conjunction with his employer in 1848 A sheet of metal cut into the form of a truncated cone is turned up and brazed, the result being, of course, a taper tube which is finished by being passed through a soft metal die. The substance of the die gives way as the tube is drawn through, but at the same time offers sufficient resistance to close the grain of the metal and give a bright surface. This must be regarded as an invention of singular ingenuity. Another method of effecting the same object is due to John Hickman, who, in 1850, patented the use of a die formed of a coil of wire, to the ends of which weights were attached. As the tube is drawn through the "die," the increasing diameter of the tube enlarges the coil, the suspended weights giving the necessary pressure.

Numerous patents have been taken out for ornamenting tubes, and for making tubes from plates which have been previously ornamented, Some of these methods are applicable also to rods, and some are strictly confined to tubes used for such purposes as the legs of bedsteads. Among the earliest we may mention the method of using engraved grooved rollers or roller-dies through which the tube or rod is to be drawn. This was patented in 1843 by Richard Prosser and Job Cutler, but for some reason or other does not seem to have come into use, for it was re-patented with some modifications in 1852 by Thomas Fearn, whose name is usually associated with this mode of ornamentation. Messrs. Peyton, Harlow, and Horne obtained a patent in 1847 for making ornamental tubes from embossed sheets, but the claim is strictly limited to the application of such tubes to bedsteads. In 1848, Joseph Gillott and John Morrison obtained a patent for producing spiral lines of ornament by causing the ornamental die to rotate as the tube or rod is being drawn through it. In some cases two dies are employed, one in advance of the other, and caused to rotate in opposite directions, by which means an ornamental chequered effect is obtained. John Cowley, of Walsall, and John Hickman, of this town, patented, in 1849, a method of obtaining ornamental tubes by perforating them, and placing variously-coloured materials inside, or by filling the interstices with porcelain, enamel, or other similar substances. Fluted taper tubes are made according to John Hickman's patent of 1850, by the use of a grooved mandrel and a corresponding series of bars arranged outside the tube, which are forced into the metal by being passed through a soft metal die.

Amongst the miscellaneous processes having reference to the manufacture of tubes, we must not omit to mention a very ingenious machine for bending tubes into a spiral or helical shape for distillers' worms and other purposes. It consists of a pair of rolls, each having a semi-circular, spiral, or helical groove cut upon them. The tube to be bent is presented to the rolls and is drawn in and coiled round one of the rolls, from which it may be removed by taking the roll out of its bearings. This was patented by Thomas Deakin in 1850.

Referring to the methods of uniting tubes of different materials by drawing, described in our last article, we omitted to mention Thomas Potts's patent of 1848, for lining copper locomotive tubes with Bath metal, for preventing the deterioration due to the rapid passage of fragments of cinder, carried through with the draught. In Messrs. Winfield and Ward's patent, above alluded to, there was a claim for drawing tubes of different metals one over the other, but it is difficult to see in what the novelty consisted. Mr. James Banister included in his patent of 1849 a method of making boiler tubes of copper, iron, and brass, the copper being outside in contact with the water, and the hard brass inside to receive the heated products of combustion from the furnace. The object of the internal iron tube was to give strength and stiffness to the whole. In 1850 Messrs. George Allen Everitt and George Glydon patented a boiler tube lined for a short distance at the end which was intended to be placed near the furnace.

## CHAPTER XIX.

### THE MANUFACTURE OF METAL TUBES.

GUN BARRELS AND GAS PIPES — ROLLING TAPER SKELPS—GROOVED ROLLS—HAMMER-WELDED TUBES —HENRY JAMES AND INTERCHANGEABLE GUNS AT TULA — A NEGLECTED GENIUS — DIMINISHING — GROOVE ROLLS — HOLLOW - FACED HAMMERS — RUSSELL'S PATENTS—SMOOTH BORE TUBES —LOP-WELDED TUBES—WELDING AND ROLLING TUBES—ROLLING BEVILLED AND SQUARE SKELPS—PROSSER'S FOUR-GROOVE ROLLS AND FIXED MANDRIL—WELDING BY SMALL GRIP HAMMERS—CUTLER'S PATENT FOR IRON OR BRASS REVERSE TUBES — MANDRIL WELDED TUBES.

DURING what may without offence be called the "hot youth" of the volunteer movement, when the superiority of the small-bore rifle for match shooting became manifest, it was customary to speak of the regulation arm as the "gas-pipe." This was a term of reproach which the "long Enfield" never merited—but let that pass. There is, however, more connection between a gun barrel and a gas-pipe than is at first sight apparent. As we have already mentioned, worn-out gun barrels were used in the early days of gas lighting for the purpose of conveying the subtle fluid, as it was then regarded, and we have heard it argued that the gasfitter's mode of expressing the diameter of a pipe as being of a certain size of "barrel," is derived from the old practice. We do not put this forward with confidence, but it may be true, nevertheless. It is, however, undoubted that the machinery for making gas and other iron tubes of a similar character grew out of the gun barrel trade. We must therefore preface this part of our subject with a few remarks on the patents for making gun barrels, nearly every one of which has originated in the town and neighbourhood.

We commence with the patent granted to John Jones, toolmaker and diesinker, of this town, in 1806, for a method of rolling skelps taper, and at the same time of a prismatic section, so that when twisted on a mandrel the coils shall lie closely against each other. This, it is evident, is not the case when a square or rectangular bar is twisted. The taper form was imparted by causing the rolls to gradually approach during the operation by means of a combination of gearing which forced down the bearings of the top roll. The inventor also describes a machine for twisting the skelp as it issues from the furnace. Benjamin Cook's method of making gun barrels by "dishing" a circular plate, patented in 1808, was mentioned in the previous article. As a substitute for the draw-bench for reducing the size of the barrel, he also proposes to employ, when desired, a series of grooved rolls. In 1811 John Bradley, of Oldswinford, patented a method of rolling gun skelps with the requisite taper, both as regards width and thickness, by using rolls having grooves of diminishing depth. In the chapter of Babbage's "Economy of Manufactures," which deals with the effect of combinations amongst workmen, the author states that this invention was due to a strike amongst the gun-skelp forgers, whose services, in consequence of the success of the machine were subsequently largely dispensed with.

There is a long story connected with the next patent on the list—that of Henry James and John Jones—which was granted in 1811. The improvements consist in welding barrels by a series of hammers, either slightly grooved on the face or plain, which strike the barrel, supported internally by a mandrel, the barrel itself being placed in a grooved swage of the proper diameter. Grooved rolls are also mentioned as an alternative method, but on the whole the inventors prefer the hammers, which are to be worked by power. The specification also includes a most ingenious self-acting lathe for turning barrels, instead of grinding them, so that the taper shall be perfectly uniform and symmetrical with regard to the bore. Jones was an extremely clever machinist, and at some period which we have not been able to fix he found his way to Russia, as many of our countrymen did about sixty or seventy years ago. He obtained employment at the Government Arsenal at Tula, which was, and perhaps is still, the Russian Enfield. The fact is very little known, but long before the "interchangeable system" was carried out here the Russian Government had perfected it at Tula, and our fellow-townsman, James, was the man who designed the machinery. About forty or fifty years ago a book was issued in Russian, illustrated by a large number of plates, giving a complete account of the establishment, and in which Jones's name is mentioned in terms of the highest praise. We saw a copy of the work thirty years ago, but all our efforts have failed to obtain exact particulars of the title, name of author, &c. Nothing appears to be known of Jones by those who have interested themselves in the history of the gun trade, and the only place in which he is mentioned is in the evidence given by the late Richard Prosser, before the Small Arms Committee in 1854. Mr. Prosser was engaged in editing a translation of the book, and many of the plates were reproduced, but his death in 1854 put a stop to the enterprise. Proofs of some of the plates and a fragment of the translation are preserved in the Patent Office Library, to which they were presented by his son. We must apologise for this long digression from the immediate subject, but no other opportunity will present itself of recording the labours of a forgotten Birmingham genius.

In 1814, George Heywood, of Brockmoor Ironworks, Oldswinford, ironmonger, took out a patent for rolling taper skelps, by the use of diminishing grooves, and also turning up skelps preparatory to welding. The latter process is effected by a projection on the top roll, and a corresponding trough-shaped groove in the lower roll. The skelps, after passing through these grooves, are completely turned up by the action of rolls, each having a semi-circular groove. They are welded in the ordinary manner.

The patents hitherto described do not, perhaps, exhibit the gradual development of tube-making machinery from that used in the manufacture of gun barrels so well as those which were granted to William Osborn, in 1812, 1813, and 1817. Osborn, we should say, was a sword and gun-barrel maker in Bordesley, and in one of his specifications he describes himself as residing at "Whitmore House, in the parish of Aston." The first patent, dated August 3, 1812, is for a method of turning up skelps by means of "vibrating swage-segments" of various sections which work in corresponding swages, or rather troughs, of the same length as the skelp. The first bends the plate into a trough-shape, the next closes it still further, and the skelp is closed ready for welding in the third. Tubes are not mentioned, though there is a hint that the machine will be in some way useful for "other articles," besides gun barrels. In his next patent, which bears date November 28 of the same year, he speaks of "cylinders of iron and steel," and also of "any other cylindrical article," although the improvement is in the main applied to the manufacture of gun barrels. The welding is performed by the action of a tilt hammer with a semi-circular groove on its face working against an anvil bearing a corresponding groove, a mandrel being used for giving internal support to the tube or gun barrel. During the short interval which elapsed between his first and second patents he appears to have found out that his machine for turning up skelps might also be applied to welding them, and he accordingly includes such an application in his patent of November. The only difference is in the form of the "swage segments." In 1813 he had another patent for rolling gun barrels, or solid rods, by passing them through rollers with grooves of a gradually diminishing diameter, an idea which, as we showed last week, has been patented again and again. Four years afterwards, that is in 1817, he patented the use of grooved rolls for welding "cylinders of various descriptions," a mandrel, with a suitable stop to prevent it being drawn through the rolls, being employed to give internal support to the tube or barrel. Tubes may be made in this way of equal thickness throughout; parallel inside and taper outside, or the reverse; or taper both inside and outside, by employing suitable combinations of taper or straight mandrels, and the rolls with tapering grooves described in his patent of 1813. It was a very important invention, though it did not come into immediate use, partly in consequence of the cheapness of hand labour and partly owing to certain mechanical difficulties. The occurrence of a strike of the barrel-forgers of the town whilst a heavy contract was in hand caused the attention of the trade to be again drawn to the machine, and it was shortly afterwards brought to perfection. This case is also cited by Babbage in his book quoted above. It is stated in the *Mechanics' Magazine*, vol. 36, that Osborn began to make gas tubes by this machine about 1820, when they were used for lighting his workshops. He also supplied Mr. Clegg, the pioneer of gas-lighting, with tubes. They were made in lengths of about four feet, but they had to be passed many times through the rolls before they were completely finished.

The patent of James Russell, of Wednesbury, dated January 19, 1824, is the first in which iron tubes for gas and other purposes are distinctly mentioned. It was of considerable importance at the time, as it produced tubes very smooth in the inside, as will be seen from what follows. The method of welding was indistinguishable from that patented by Osborn, consisting as it did merely of a tilt hammer with a semi-circular groove worked in connection with a corresponding grooved bolster or bed. When the welding operation was completed, the tube was again heated and passed through grooved rollers, a bulb-shaped mandrel being firmly fixed in the axis of the tube, so that immediately the latter had left the grip of the rolls it was forced on to the mandrel. This had the effect of clearing and smoothing the inside, and producing for the first time a tube capable of resisting the corrosive effect of the gas of that period.

The patent which laid the foundation of the present welded iron tube trade was, however, that granted in 1825 to Cornelius Whitehouse, a workman in the employ of James Russell, of Wednesbury. [There was a Cornelius Whitehouse who carried on the business of a jobbing smith at 78, Lichfield Street, Birmingham, about the year 1770. Could he have been a relative?] The turned-up skelp having been heated in a furnace, was welded by drawing it, without a mandrel, through a pair of semi-circular dies, or through pincers, each jaw of which had a semi-circular groove or aperture. The patent created a revolution in the trade, the price of tubes was reduced to two-thirds, or, in some cases, to half, and much greater lengths could be made. They were of greater uniformity, both internally and externally, and the trade came very soon almost entirely into the hands of Russell, to whom the patent was assigned. As a matter of course it was infringed, and the litigation which commenced in 1830 was not finally set at rest until 1848, when the great suit of Russell *v.* Ledsam was concluded in the House of Lords, long after the patent had expired. In all the actions which Russell commenced the patent was upheld, and in 1839 it was extended for six years, or until 1845. Up to the time of the grant of the prolongation Russell's law expenses had amounted to upwards of £4,000. It was one of the most obstinately fought patent cases on record, but several of the points were in the highest degree technical. The validity of the extension was disputed on the ground of the application not having been made in time. Then the Lords of the Privy Council made it a condition of the extension that the original inventor, Cornelius Whitehouse, should have an annuity of £500. a year secured to him. It was contended by the defendants in Russell *v*. Ledsam that the Crown had no power to insert such a condition, and they alleged further that the annuity had not been duly paid. Amongst the testimony to the usefulness of the invention was that of Perkins, the patentee of a well-known system of warming buildings, who stated that no other tubes would have enabled him to carry out his invention. They would bear bending cold, and would stand a pressure of 5,000lb. on the square inch. Francis Bramah, son of the great inventor, also stated that the introduction of the improved pipe enabled him to dispense with the very expensive copper pipe formerly used, and which cost 10s. per foot, whilst Russell's tubing could be obtained of sufficient strength at 1s. 3d. per foot.

The patent of George Royl, of Walsall, dated 1831, consists in a method of welding tubes by passing them through a pair of grooved rolls placed close to the mouth of the furnace used for heating the turned-up skelp. The tube was afterwards passed through a "scraper," but the courts decided, in Russell *v*. Cowley, that grooved rolls were a mere mechanical equivalent for the dies of Whitehouse's patent; and it was further contended that the scraper or "scorpion," as it was also called, was the organ which really performed the operation of welding, the rolls being merely a colourable addition. The question of the similarity of the process of welding by drawing a tube through dies and rolling it through grooved rolls was one which occupied some of the most eminent counsel of the day, besides engaging the attention of at least half a dozen of the judges for a very long time. In 1836, Frederick Edward Harvey, a draughtsman employed at the Horseley Iron Works and Jeremiah Brown, a Tipton roll-turner, obtained a patent for rolling skelps with bevelled or square edges, according to whether a "lap" or "jump" joint was required. The skelp was turned up by first bending the ends for a short distance by rolls having eccentric and peculiarly-formed grooves, the operation being completed by passing it through grooved rolls with the assistance of a stationary mandrel. The welding operation was effected by grooved rolls, which forced the heated tube on a fixed mandrel, the bulb or projecting part of which was placed between the rolls at the point where the grooves meet. Although the Thomas Henry Russell to whom a patent was granted in 1836 is described as of Took's Court, Chancery Lane, he was a member of the Staffordshire family of that name, who have been so long connected with the trade. The improvement consisted in turning up and welding a skelp at a single heat, the first-named operation being performed by grooved rolls in combination with a revolving plate projecting into the opening formed by the edges of the skelp, the object of which was to keep the seam perfectly straight. From the rolls it passed through welding dies in the ordinary manner.

The principal feature of Richard Prosser's patent of 1840 was the use of four grooved rolls and a fixed mandrel with a bulb on the end. It was found, when using a single pair of rolls such as have been several times mentioned in the course of this and the previous article, that the pressure was very unequally distributed round the circumference of the pipe. Mr. Prosser proposed to overcome this difficulty by combining two pairs of rolls at right angles to each other, so that when one pair are operating to compress the metal vertically, the others are acting in a lateral direction. The inventor also describes a machine in which three rolls are used. Two or more welding machines may, when desired, be arranged close together in line, the rolls of the second being slightly smaller than those of the first, so as to further compress the tube, which is passed through the entire set at a single heat. By arranging the machines in a curve bent pipes of any required radius may be obtained. The specification also includes a reverberatory furnace for heating the skelps. Lengths of ten feet could be made by this machine, and the action was very rapid, a period of two seconds only being necessary to weld a tube of that size.

This patent was assigned to Mr. Charles Palmer, who arranged to pay the patentee a royalty of £4. 13s. 4d. for every ton of pipes manufactured. Palmer assigned it to Bower in 1842, and subsequently George Selby and W. R. Hodges became possessed of shares in the patent. The two latter started the "Patent Tube Company," at Smethwick, in 1845, since which time the process has been carried on uninterruptedly, some of the original machines being still in use. In the same year the inventor, in conjunction with Job Cutler, who was then residing in Ladypool Lane, commenced a tube manufactory at Coatbridge, near Glasgow, which they called the "Caledonian Tube Company," and this manufactory is still in existence. A great deal of litigation took place between the parties, and it would almost appear that in those days any one who became connected with the tube trade was certain to be involved in a tedious lawsuit before very long. About the year 1847 proceedings were commenced by the late Irving Van Wart—his son Henry and S. A. Goddard were also interested—to set aside the patent by *scire facias*, but we believe that they were not successful. Some curious particulars of the rise of the iron tube trade of the town are contained in the proceedings in Regina *v.* Cutler, Regina *v.* Prosser, and Bower *v.* Hodges and Selby, which may easily be found by a reference to the law reports of the day. There were also one or two lawsuits in the Scotch courts.

Job Cutler's patent of 1841 was for welding tubes by drawing them through a die, or grooved rolls, using a mandrel to give internal support during the operation. Another mode was by the employment of a number of small hammers working with great rapidity, which were caused to act upon the seam of the tube as it was slowly drawn from the heating furnace. A mandrel was inserted inside the tube to serve as an anvil for the hammers to act against. There was also a claim for covering iron tubes with copper or brass, and applying them in the construction of tubular boilers ; and the inventor describes a method of drawing tubes cold, so as to give them a dense and uniform surface. As might have been expected proceedings were instituted to cancel this patent by *scire facias*, on account of its want of novelty, and because it infringed existing patents.

In 1842 a patent was granted to Thomas Henry Russell and Cornelius Whitehouse for a method of welding tubes when a mandrel was used. The mandrel employed was smaller than the diameter of the tube, so that when the pressure was put on, the tube was squeezed into the form of an oval, the shorter diameter of which was the same as the diameter of the mandrel. To release the mandrel the tube is again heated and passed through cylindrical dies or rollers, which by altering the shape sets the mandrel free. Reference has been already made to the extraordinary amount of legal acumen which was displayed upon the question of the identity of action of dies and grooved rolls. By an ingenious use of language the patentees affirm, or appear to affirm, that grooved rolls are mentioned in Whitehouse's former patent. They say in the document under consideration, "We do not confine ourselves to the use of dies in the form of tongs, as grooved rollers or other dies, such, for instance, as are described in Whitehouse's patent of 1825, may be used." There is not a word about grooved rolls in that patent !

## CHAPTER XX.

### THE MANUFACTURE OF METAL TUBES.

(CONCLUDED.)

ROOSE'S PATENT—WELDED TWIST BARRELS—HARDY'S TREBLE ROLLS—RUSSELL'S CARRIAGE TUBE—SELBY'S PATENT—PROSSER'S SPLIT EDGE WELD—WHITEHOUSE'S ROTARY HAMMER—DR. CHURCH'S CONICAL TUBES—PROSSER'S ANTI-WELDED TUBES AND TUBE EXPANDER.

With the exception of a mode of turning up skelps, there is nothing relating to the actual formation of tubes in Prosser and Cutler's patent of 1843, the improvements having reference for the most part to the methods of treatment after manufacture. It includes a mode of forming sockets on iron pipes by the action of a plunger which is forced into the end of the pipe when heated. The external form of the socket is produced by a die, into which the tube is inserted during the operation. Two methods of removing excrescences from tubes or bars by revolving cutters, scrapers, or rolls are described, as also an arrangement of revolving brushes for freeing tubes from scale in the welding machine. The claim for ornamenting, embossing, fluting, or reeding tubes by drawing them through engraved rolls was referred to in a previous article.

In 1843 a patent was granted to James Roose, formerly of Birmingham, but then of West Bromwich, for a method of supporting and stiffening the mandrel used when welding tubes by drawing through holes. A claim was also made to the construction of the tongs used for grasping the tube. In the same year Roose had another patent for welding twisted gun barrels and cannons, one feature of which was a method of pressing the coils into close contact during the operation. Mr. James Hardy, of this town, who was the inventor of a most valuable method of forging railway axles, took out a patent in 1844 for arranging welding rolls "three high," as it is now called. When the tube had been passed through the lower set it was immediately lifted up by self-acting gear and put through the upper pair, so that no time was lost, and the welding was completed at a single heat. In the following year he took out a patent for the application of a well-known arrangement of reversing gear to tube-welding rolls, with the same object in view. John James Russell and Thomas Henry Russell, of Wednesbury, had a patent in 1844 for a method of welding, in which the tube was mounted on a carriage or travelling bed which was drawn under a grooved welding roller, the tube being supported internally by a mandrel. The machine was specially applicable to the manufacture of tubes of large diameter or considerable weight. The elaborate specification of the patent granted in 1845 to Mr. George Selby—then of the Tube Works in Cambridge Street, but afterwards of Smethwick, as already stated—contains a machine for turning up skelps, consisting of two pairs of rolls, one in front of the other. The pair nearest the furnace bend the heated skelp into a gutter shape, and it passes on from these to the second pair, which completely round it. The welding is performed in Prosser's four-roller machine, to which scrapers are attached to free the tube from scoria just as it enters the welding rolls, and cutters to remove the "fins" or ribs left on the tube by the junctions of the rolls. There are also improvements in the mode of arranging the welding furnaces, and in the mouth-pieces for guiding the tubes into the grooved rolls. Tubes are straightened by stretching them when hot, aud cut to length by a machine of the nature of a lathe with two hollow mandrels, both ends being cut off at the same time. George Royle's patent of 1845 is almost exactly the same as that granted to him in 1831, and referred to above. It consists of a method of welding pipes by a combination of grooved rolls and dies.

A new principle of tube making appears in Richard Prosser's patent of 1845, although it was never successful on a large scale. The edges of the skelps were fashioned in such a manner that when turned up they should not only meet but enter the one into the other. To effect this, a V groove was cut in one edge, whilst a "double bevel" was cut on the other edge, or both sides were "rabbeted." This was effected by drawing the skelp against cutters. The turning up was done by moulds, or dies as they may be called, the whole length of the tube, which were made to approach each other by a hydraulic press. The first pair caused the skelp to assume a gutter shape, the second advanced it a stage nearer to the circular form, whilst the third pair closed the tube completely, the edges fitting into each other as before described. It was said that a much stronger weld could be produced when tubes were prepared in this manner. The specification is of very great length, and contains many improvements in the details of the tube manufacture. There is a carriage for transferring tubes from the furnace to the welding machine, and an ingenious tool for cutting grooves in the welding rolls, so that a perfect circle shall be produced at their point of contact. The welding machines are to be provided with wheels running on rails, so that they may readily be removed for repairs; and a method of cleaning and finishing tubes is described, according to which they are placed in an iron barrel kept slowly revolving. In this manner each tube is cleaned by friction against its neighbour, and the inside may be acted upon by placing a loose iron rod in the tubes. A self-acting machine for cutting tubes to length is also mentioned, as well as a collapsible mandrel, to prevent breakage of the machinery.

In 1845, Cornelius Whitehouse, who then described himself as of Wolverhampton and Birmingham, gun-barrel manufacturer, obtained a patent for a method of welding gun barrels and tubes by what he called a "rotatory hammer," but which was in fact nothing more than an ordinary roller, the periphery of which was discontinuous. It was thought that if portions of the circumference were cut out so as to produce, for instance, a roller of a star shape, that a hammering effect, similar to that due to a succession of blows, would be produced. Twisted gun-barrels were welded in a sort of die, made to fit the barrel, the end pressure necessary to cause the coils to adhere being given by a screw. There was also a claim for equalising tubes by drawing them over a mandrel, part of which was of larger diameter than the tube is to be finished at, and causing the tube to be reduced to a smaller diameter as it is drawn off the mandrel. Thomas Henry Russell's patent of 1845 consisted in the use of a "beak iron" for giving internal support, instead of the ordinary mandrel, the tube being drawn under a welding roller by means of the ordinary draw-bench.

Dr. Church's patent of 1846 embraces a variety of subjects, but the part relating to the manufacture of tubes consists of a modification of Prosser's method of turning up skelps by the use of long moulds, the main difference being that a mandrel is employed. He also proposes to weld tubes in a similar manner, and by varying the contours of the moulds conical tubes may be produced, and, indeed, almost any hollow article, such, for instance, as hollow rails or beams. An attempt was made by James Roose, in 1840, to turn up and weld a skelp at a single heat by drawing it from the furnace through a trumpet-mouth and then through welding rolls, the end having been previously bent over in the usual manner. A blast of air is directed on to the seam during the weld so as to keep up the heat. The object of Cutler and Robinson's patent of 1848 was to make boiler tubes thicker at one end than the other, which was accomplished by rolling them, at a certain stage of manufacture, with a taper mandrel inside.

The metal was thus gathered, as it were, towards one end, and in fixing the tubes in the boiler the thick end was placed near the furnace, so as better to resist the great heat of the fire. The mandrel was removed by a draw-bench, or by "cross-rolling."

Great things were expected from the "anti-welded tube" which was introduced into the market in 1850, but it failed to satisfy the requirements of the public, and it disappointed the hopes of the inventor, the late Richard Prosser, who spent upwards of £20,000. in bringing it out. As its name imports there was no weld, the edges of the skelp, which had been previously accurately "rabbeted" in a planing machine, were simply turned up so as to meet in close contact. The invention really grew out of Prosser's former patent, and the anti-welded tubes may be regarded as tubes made according to that patent, but not welded. They were obviously only suitable where the pressure was external, as any force from the inside would tend to open the seams and cause the tube to leak. They were tried in a locomotive engine on the London and North-Western Railway, but they were unsuccessful. Not only did the tubes require careful handling so as to avoid accidental injury, but the seams were liable to be opened during the operation of removing obstructions. Eventually the whole of the stock was sent to be welded in the ordinary manner. The machinery employed in the manufacture was of a very elaborate and ingenious character. The skelps were planed at one pass through the machine by means of a compound cutting-tool having several distinct cutting edges acting in succession. The skelps were turned up in the same manner as that described in the patent of 1845, except that a "steam hydraulic" press was used. Since that time the principle has received numerous applications, but the name of the inventor has been forgotten. The ordinary force pump is dispensed with, water under pressure being supplied to the hydraulic press by the action of a plunger set in motion by a piston working in a large steam cylinder. When the machinery at Mr. Prosser's works in Cambridge Street was sold in 1854, the press—the first of its kind—was bought by a Mr. Smith, a manufacturer of railway wheels at Smethwick. The specification also contained a description of an hydraulic riveter, and of an elastic ferrule for fixing tubes in boilers, which differed from those previously in use by being driven between the tube and the hole in the tube sheet, instead of inside the tube.

Before bringing this article to a conclusion we may put on record the fact that the tube expander for fastening boiler tubes in their places was also the invention of the late Richard Prosser, who manufactured large quantities of them. It was not patented in this country, but was included in his Scotch patent for making tubes granted in 1845. It consisted of a number of hardened steel dies kept loosely together by an elastic ring. In the centre of the dies is a hexagonal hole through which a correspondingly-shaped plug passes. The dies being placed in the tube to be expanded, the plug is drawn forwards by a powerful screw, the effect of which is to force out the dies in a radial direction. The pressure which they exercise on the tube in this manner causes it to be forced into close contact with the hole in the tube plate. This particular form of expander has now been superseded, but for many years it was the only efficient substitute for the old method of tightening the ends of tubes by driving in a taper plug.

## CHAPTER XXI.

### LOCOMOTIVE STEAM CARRIAGES.

EARLY RAILWAY PROPOSALS—WILLIAM JAMES AND STEAM CARRIAGES—HEATON'S STEAM CARRIAGE—ITS LOCAL JOURNEYS—DR. CHURCH'S STEAM COACH — THE "STEAM CARRIAGE COMPANY" — DR. CHURCH'S LOCOMOTIVE "ECLIPSE."

MOST of the preceding articles have been devoted to branches of trade or manufacture which have either originated in the town or have taken root there as in their natural soil. But steam locomotion did not take its rise here, nor have the labours of our townsmen exercised any abiding influence upon the subject at large. We cannot put forward a claim to compete with the more northern towns for the honour of having rocked the cradle of the locomotive; but something was done in Birmingham, nevertheless. Much ingenuity was displayed and much money was spent, and we were early in the field. It ought not to be forgotten, as it too frequently is, that a Warwickshire man was one of the pioneers of the present railway system. Out of every thousand persons who can give an intelligent account of George Stephenson's labours how very few know anything whatever of William James! This is not the place to discuss the rival claims of these two men, and we need do no more than briefly state that James was born at Henley-in-Arden on the 13th of June, 1771, and that he was an earnest advocate for railways, long before Stephenson appeared in the field, many of the surveys for the earlier railways having been made by him. He was brought up as a solicitor, to which profession he added that of a land agent, surveyor, and engineer. James's claims are very fairly stated in a little book by "E.M.S.P.," entitled "The Two James's and the Two Stephenson's," published in London, in 1861. It should be looked after by local collectors, as it contains much information relating to this neighbourhood. James's eldest son, William Henry, was born at Henley in 1797, and he appears to have inherited much of the talent of his father. His education was confided in the first instance to the Rev. Dr. Smith, head master of King Edward's School, and the little work above mentioned contains many details of his early inventions. In 1824 we find him established as an engineer at Coburg Place, Winson Green, where he resided for fourteen years. In 1838 he removed to London, but his career was unfortunately not prosperous, and for many years he was incapacitated by ill-health. In 1867 an appeal was inserted in the *Engineer* newspaper on his behalf, stating that he was in a state of destitution. He took out twelve patents relating to various subjects, including steam carriages, railways, diving apparatus, telegraphs, weighing machines, and heating and cooling fluids. His first patent, granted in 1824, was for a steam carriage, the driving wheels of which were actuated by an independent engine. This gave great facilities for turning corners, and the steering apparatus was connected in a very ingenious manner with the steam valves, so that, in nautical parlance, the vehicle must have "answered its helm" with great rapidity. In 1832 he took out another patent for a steam carriage of a much more complicated construction, but we are not aware that any trials were ever made in Birmingham. The boiler used was very effective; but as he is said to have invented the multitubular locomotive boiler, it is well to say that none of his specifications bear out the assertion that he anticipated Stephenson in his application of tubes to the "Rocket," in 1829.

The next machine of the kind which originated in the town was that patented in 1830 by Messrs. John, William, George. and Reuben Heaton, sons of Ralph Heaton I., and uncles of the present Mr. Alderman Ralph Heaton, who is the third of that name, his eldest son being the fourth. It was not exactly a "steam carriage," but rather what we should now call a "traction engine," as it was only adapted for drawing vehicles, not for carrying passengers. In fact the inventors called it a "steam drag," which sufficiently expresses its object. One of its earliest appearances in the streets of Birmingham took place in October, 1830, and a Bilston correspondent of the *Mechanics' Magazine* says that he was much "surprised"—as indeed he well might have been—on the 13th day of that month, "to meet a locomotive steam-engine in Bath Street, drawing a large cast-iron press, which, on enquiry, I was informed, weighed one ton eleven cwt. exclusive of the carriage on which it was placed. It proceeded at a steady pace up Whittall Street, Steelhouse Lane, Bull Street, Temple Row, Waterloo Street, and Ann Street, to the manufactory of Mr. Hardman, button and medal manufacturer, where the load was detached, and the engine started off at an amazing velocity, turning the corners of the streets right and left with the greatest exactness." The name of the engine was the "Enterprise." Nothing seems to be recorded of the machine for about three years, but in the *Magazine* for June 6, 1833, there is a letter from Messrs. Heaton describing the transport of a heavy press to Dudley. On that occasion nine miles were accomplished, with a load of 25 cwt., exclusive of the carriage and chains and five persons, in 2 hours and 8 minutes, 38 minutes of that time having been spent in procuring water and adjusting the load, which had accidentally slipped. On the 16th of the next month the energetic brothers made a successful trip to Wolverhampton, a light wagon containing 22 persons being conveyed thirteen miles in 1 hour and 56 minutes, 16 minutes of which were occupied in taking up and setting down passengers and taking in water. The return trip was not so successful, two accidents having happened to the machine during the journey. The engine was out again as soon as the damage was repaired, and at 6 minutes past 10, on the 28th of August, it started from the mills in Shadwell Street, with a stage coach carrying fifteen passengers. Five others were picked up on the road ; and the Bell Inn, Northfield, seven miles distant, was reached in four minutes under the hour. The Lickey Hill, which is nearly half a mile long, and rises on an average 1 in 9, was successfully ascended, the coach and nine passengers being conveyed to the top in 9 minutes. From thence they proceeded to Bromsgrove, and on the return journey—to show how perfectly the machine was under control—the conductors brought it to a stand on the steepest part of the hill. The manufactory was reached at 35 minutes past 4, the sharp ascent of Worcester Street, which rises 1 in 12, being surmounted without difficulty, although the coach was at the time loaded with 32 passengers. The coke used during the entire journey of 29 miles was 11 bushels, the value of which was half-a-crown. These particulars are taken from a report which Messrs. Heaton communicated to the *Mechanics' Magazine* of September 14, 1833.

Encouraged by their success, steps were taken to form a company, and a meeting was held at the Union Tavern, in the middle of October, to appoint a committee. It was originally intended to issue 2,000 shares only, at £10. each, but in less than 24 hours 3,000 were eagerly purchased. By the following April they had risen to 10 prem., but the scheme was not a success, as the expenses of wear and tear attendant upon a speed of ten miles per hour were found to be too great to allow of profitable running.

Meanwhile another competitor had appeared in the field, and public opinion was much divided as to the relative merits of Heaton's carriage and that brought out by the indefatigable schemer, Dr. Church, who obtained a patent for a steam coach in February, 1832. What the real thing was we have no means of knowing, but according to a print published by Josiah Allen at the time, it was a sort of cross between the car of Juggernaut and a French diligence, with the additional peculiarity of having only three wheels. To the eye of the mechanic it presented a heavy, lumbering, and decidedly top-heavy appearance. Passengers were carried in a sort of hooded seat in front, similar to the *banquette*, as we believe it is called, of a diligence, on the roof, and in a *coupé* in front and behind. The centre of the structure is given up to the machinery and boiler. The drawing attached to the specification shows a somewhat different arrangement, there being no provision for outside passengers. The framing is specially designed to obtain as much stiffness as possible. The driving-wheels, which are enormous, are worked by pitch-chains from the engine shaft, and are supported on air springs. The boiler is upright, with a peculiar arrangement of syphon-tubes through which the heat passes. The engines are direct-acting, on the same construction as those for which the inventor obtained a patent in 1830, but in the steam-carriage their position is reversed, the crank shaft being below the cylinder. In 1835 Dr. Church obtained another patent, which contained several other and far handier forms of steam carriage. The document is of great length, and refers to a variety of subjects connected with the steam-engine. It is illustrated by no less than 21 large sheets of drawings.

For the purpose of working the patent the "London and Birmingham Steam Carriage Company," was formed, and it was stated in the prospectus—issued about June, 1832—that the engines would be of 60-horse power, and capable of conveying a weight of 15 tons at a rate of 15 miles per hour. A certain "J. H. B.," who wrote several letters dated from Birmingham in the *Mechanics' Magazine* of the day, warns the public "not to be entrapped into a delusive scheme," and twits the company with never having made any experiments to prove the truth of their assertions. Another correspondent advises the Doctor to "clip the wings of his flighty imagination, descend to more moderate plans, and confine his extravagant notions within more rational dimensions, before his attempts can have any chance of realising that practical success and utility which I wish him from the bottom of my heart most sincerely." In spite of this, and certain facetious remarks of another writer who called himself "Junius Redivivus," the shares continued to go up, and in April, 1834, they were at 20 to 25 prem. The carriage was first put upon the road by way of experiment at the latter end of April, 1834, but it met with an accident in turning a corner, and was hauled back by the workmen. We hear of it again through the local papers of the day in October of the same year. On the 6th of that month a trip was made on the London road as far as Stone Bridge, when it attained a speed of ten miles per hour, even when ascending hills. An accident happened on the return journey which rendered it necessary to blow off the steam and draw the fire. When the company was finally wound up we are unable to say, but in April, 1835, the *Mechanics' Magazine* taunts the directors with their delay in putting the carriages on the road. It was announced about that time that the directors were authorised to build a sufficient number to establish a regular system of communication between Birmingham and the metropolis. We believe that only one carriage was actually built, and it would be very interesting to know what became of it. For many years it lay disabled by the roadside, in the neighbourhood of the Green Lanes, but it has probably long ago disappeared in the "infinite azure" of the melting pot. There was a Mr. John Huggins, who was with Dr. Church at the time. who might probably throw some light on the matter. Mr. Huggins is, we believe, still residing somewhere in the neighbourhood.

In spite of the failure of these attempts the notion that steam locomotion on common roads would eventually succeed appears to have survived in the breasts of the enthusiasts for some years. As late as November, 1837, a Birmingham correspondent of the *Mechanics' Magazine*, signing himself "L. M. B.," enquired as to the prospects of the various schemes which had been proposed. This brought forward Mr. George Millichap, an axle-tree manufacturer, of 85, Aston Street, who stated that he had a steam carriage nearly ready for trial. It was probably upon the principle of that mentioned in his patent of 1834, which consisted in a mere arrangement of gearing for actuating the wheels, and to which a steam engine *might* be applied. Millichap was also a patentee of axle-trees and boxes in 1820 and 1844. In his patent of 1820 he describes himself as of Worcester, coach builder.

Although it does not strictly belong to our subject we may here put on record the fact—vouched for by Mr. F. W. Turner, of Swansea, in a letter to the *Engineer* of June 14, 1861—that Dr. Church's locomotive "Eclipse" was sold to haul coals on the Swansea Vale Railway. The engine was built in 1838, and was the first which ran a mile a minute, and was also the first tank-engine on four wheels. The construction was marked by profound mechanical skill; the connecting rods were hollow, and not parallel, but tapering according to the correct proportions. The reversing gear was elaborate and the piston well designed. In the following number of the above-mentioned journal Mr. Richard Bach, of this town, wrote to say that the boiler which was in the "Eclipse" was not the original one, which exploded shortly after the engine was set to work, and was then refitted with a boiler of another construction.

It was remarked at the outset of this article that the experiments with steam carriages in this town did not lead to any important practical results, but an exception must be made in favour of Mr. George Heaton. During their early trips it was found that the engine was so unsteady when working at 160 or 180 strokes per minute, that it was impossible for the men to keep their seats. "Being aware," Messrs. Heaton observe, "that this motion could only be produced by some portions of our machinery being out of balance, we placed a compensating weight opposite each crank, and repeated our former experiment upon the same road, and found we attained greater speed with no greater consumption of fuel, and the machine travelled perfectly steady at any speed, and free from any symptoms of rocking or shaking." In due time the principle was applied to the counter-weighting of locomotives, with the most satisfactory results. Mr. George Heaton still continued to pursue the subject, directing his attention to the compensation of the disturbing action of the *reciprocating* weights in an engine. In 1847 he accordingly took out a patent for counterbalancing pistons by causing a corresponding weight to move in an opposite direction. It is not necessary to say more on the subject, as all the facts are on record, and it was, moreover, thoroughly discussed in the *Engineer* so recently as January 30 last.

## CHAPTER XXII.

### ELECTRO-METALLURGY.

EARLY EXPERIMENTS — WOOLRICH'S DEPOSITS BY VOLTAIC BATTERY—ELKINGTON'S PATENT FOR GILDING—SHORE'S PATENT FOR DEPOSITING COPPER OR NICKEL—WRIGHT'S DISCOVERY OF ELECTRO-METALLURGY—PATENTED BY ELKINGTONS—DEVELOPMENT OF THE PROCESS — MAGNETO-ELECTRO PROCESS — WOOLRICH'S PATENT—DEPOSITION OF BRIGHT COATING—MILLWARD AND LYONS' PATENT—S. B. SMITH'S PATENT.

THE history of the origin of electro-metallurgy has been written so often that it is not necessary to enter upon the delicate question as to the rival claims of Jacobi, Spencer, and Jordan, all of whom have their partisans, and whose pretensions have been discussed with the warmth usually attributed to rival theologians. The subject attracted the attention of scientific men in this town at a very early period, and one of the earliest manuals on electro-metallurgy was that published by Mr. George Shaw, in 1842. Without the least intention of re-opening a controversy now happily closed, we feel bound to quote the following passage from the above work, of which a second and much enlarged edition appeared in 1844. Mr. Shaw says:—"It has been demonstrated to the author of this treatise, on the most conclusive evidence, both documentary and personal, that the late Mr. Woolrich, of Birmingham, succeeded in applying coatings of copper to various articles by means of the voltaic battery, and that in one instance, at least, he succeeded in gilding an article by voltaic precipitation from the ammoniacal solution of gold, before the published experiments of Mr. Jordan and Mr. Spencer appeared. Mr. Woolrich, however, never published an account of his experiments, as from the imperfect nature of the voltaic batteries then in use the deposition of copper was considered too troublesome to be valuable; and from the same cause he was unable to repeat his experiments on gilding."

Before dealing with the electro-metallurgical patents, properly so-called, we may premise that for several years Messrs. Elkington, who originally carried on the business of gilt toy makers, had directed their attention to the subject of gilding and silvering metals. In 1836, Mr. George Richards Elkington obtained a patent for gilding by dipping in a solution of gold in aqua regia to which bicarbonate of potassium had been added. The method was said by the inventor to be an efficient substitute for the old mercurial process. In the following year his brother Henry patented a mode of gilding in which the articles were first covered with a coat of platinum or palladium. The specification also included a process for gilding by means of a solution of gold and mercury. In 1838, Mr. G. R. Elkington, in conjunction with Mr. Oglethorpe Wakelin Barratt—who is variously described in his several patents as a manufacturer, a metal gilder, and an experimental chemist—took out a patent for coating metals with zinc in the wet way, as opposed to the method of galvanising. There was also a plan for colouring iron and steel to imitate brass, which consisted in coating the articles first with copper, then with zinc, and afterwards submitting them to the action of heat in a closed oven. A partial fusion took place, the result of which was the formation of a thin coating of brass upon the article under treatment.

No battery was used in any of these processes, and the first Birmingham patent in which electricity was employed was that obtained by Mr. Joseph Shore, of this town, in March, 1840, the claim being for the deposition of copper or nickel upon metallic surfaces by the voltaic battery, either for the prevention of rust, or for obtaining an improved surface. The specification was very carefully drawn so as not to include the electrotype process which had been recently discovered. The details of the method need not be described, but it should be mentioned that the idea of using nickel as a preservative coating for iron has been revived of late years, and is now largely carried on—especially in America—as a separate branch of trade.

But the grand discovery in electro-metallurgy was made by Mr. John Wright, a surgeon, who resided in the old-fashioned house in Bordesley, opposite Peyton's bedstead mânufactory, and a few doors from the Great Western Railway Station. Wright was an excellent chemist and very fond of experimenting. In order to make the precise nature of his invention quite clear, it is necessary to say that solutions of nearly all the salts of silver and gold are reduced when submitted to the action of an electric current. This was known as a scientific fact, but the deposit obtained—though perfectly pure—was not sufficiently firm and coherent to be of any use, nor was the colour good. Many solutions had been tried, but without result, until Wright happened to come across a passage in Scheele's "Chemical Essays" where the solubility of the cyanides of gold and silver in cyanide of potassium is mentioned. The passage is given at length by Dr. Percy, in his "Metallurgy of Silver and Gold." It immediately occurred to him to try this particular compound, and he succeeded. Unfortunately for his fame he did not obtain protection for his discovery in his own name, the patent having been taken out by Messrs. Elkington, in March, 1840. There is no reason, however, to think that he suffered pecuniarily, as the patentees undertook to pay him a royalty of one shilling for every ounce of silver or gold deposited. The inventor did not live long to enjoy the fruits of his labours, his death having taken place in 1844, but the firm entered into an arrangement with his widow to pay her a handsome annuity. The lady is still living, and is well known in Sheffield as Mrs. Cammell, widow of the late Mr. Charles Cammell, a noted steel manufacturer. Such is a bare outline of the history of a discovery—or perhaps we ought to say invention—which is of enormous industrial importance. Meagre as the facts are, they are not known so widely as they ought to be, and many writers on the subject appear to be totally unacquainted with the name of the "first and true inventor" of electro-plating. Even the jury at the Exhibition of 1851 were so ignorant of the origin of the art that they referred in their report to Messrs. Elkington as being the introducers of the process into this country. This want of knowledge is no doubt due to omission of Wright's name from the patent, and also to the fact of his early death. He never published any account of his experiments, so far as we are aware, and at this distance of time it is hardly to be expected that the details could be recovered. If there is any portrait of Mr. Wright in existence, we hope it will eventually find a place in the new Art Gallery amongst the "Birmingham Worthies," as proposed by "Este," in the *Daily Post* last week.

As is the case with all great inventions, considerable difficulties were experienced in introducing it commercially, and some years elapsed before electro-plating could be said to "pay," taking into account, of course, the preliminary outlay in experiments and so on. The silversmiths and platers were utterly opposed to it, and would have nothing to do with electro-plated goods, but the excellence of the products of the new art was indisputable, and in 1842 the demand was sufficient to give employment to 300 operatives at Messrs. Elkington's establishment. Of course there were prejudices against electro-plate as there have been against everything else, but it is curious that a body like the jury of the Exhibition of 1851 should have shared them. In their report they "desire to guard against being considered as expressing an opinion on the merit of the application of the electro process of *silver* plating to objects of domestic use. They desire only to commend the artistic application of this discovery, to which alone they are inclined to think it adapted. At the same time they acknowledge that the application of *gold* by this process is a highly meritorious invention, tending alike to the economy and durability of the metal applied, and to preserve the health of the artisan from the dangerous emanation of quicksilver vapours." There is a good, old-fashioned, conservative ring about all this, and for once the jury seem scarcely to have risen above the average intelligence of the conventional "twelve men in a box." They gave utterance to similar sentiments when speaking of the exhibit of Messrs. Christofle, the French licensees of the patent, and who, by the way, were compelled to spend large sums to protect their rights. It does not fall within our province to pursue this part of the question, but those who desire to do so will find a full account of the French litigation in the *Mechanics' Magazine* for February 10, 1844, and July 5, 1845.

Amongst the brilliant discoveries with which Faraday's name is associated, that of magneto-electricity assuredly stands in the front rank. He found that when the pole of a magnet is made to approach towards and recede from the core of a coil of insulated copper wire, alternating shocks are induced in the coil. These alternate shocks may be made to act in the same direction; they can then do the same kind of work that is accomplished by a galvanic battery. The idea of applying the currents so developed to the purposes of electro-metallurgy belongs to Mr. Woolrich, who patented a machine for producing them in such a way as to be applicable to electro-deposition, in August, 1842. The patent was worked by Mr. Thomas Prime, of Northwood Street, and Dr. Percy, in the work above referred to, gives a very interesting account of Faraday's visit to see the process. Dr. Percy says:—"In 1845 I conducted Mr. and Mrs. Faraday to Mr. Prime's works, where for the first time that great philosopher saw his discovery of the magneto-electric current applied to the electro-deposition of silver. I shall never forget the sparkling delight which he manifested on seeing this result of his purely scientific labours rendered subservient to a beautiful art and to the advantage of others." The superior constancy of the current obtained by the magneto-electric machine over that furnished by the battery was soon manifest, and it is now almost, but not quite, in universal use as a source of electricity for coating metals. There was another novelty in Woolrich's patent, which consisted in the application of the double sulphites of silver, gold, and copper, with the alkalies as depositing solutions. But for this improvement the patent would have been useless by reason of the previous patents of Elkington, and of Dr. Leeson, a professional man whose assistance had been obtained by Elkington to carry out the new process. Leeson claimed the use of no less than 480 salts for depositing silver, Woolrich's specification contains full instruction for preparing the various salts according as it is desired to deposit silver, gold, or copper, but it is not necessary to enter into details here, especially as this part of the patent is not now used. Dr. Percy has in his laboratory a sample of the silvering solution made by Mr. Woolrich himself in the year 1845, from which a lustrous coating of metallic silver has been deposited on the internal surface of the bottle in which it is contained. According to Mr. Woolrich's statement such deposition takes place on a surface of glass or earthenware but not upon wood, and it is independent of the action of light. Woolrich's patent was eventually purchased by the Elkingtons.

The deposits obtained by the cyanide process had one fault—they were not bright, but had a frosted appearance, and required to be burnished. The method of obtaining a bright coating was discovered by accident in a very curious way. The following is Mr. Gore's account of the discovery : "Some wax moulds, partly saturated with bisulphide of carbon, were occasionally immersed in the silver-plating vats at Messrs. Elkington's establishment, for the purpose of receiving a coating of silver ; and it was observed by Mr. Millward, the silver depositor, that whenever these moulds were in the vat the silver deposited on the articles near them acquired a bright appearance. This led to experiments upon the influence of adding small quantities of bisulphide of carbon to the ordinary silver-plating solution, and it was soon found that this substance produced the desired effect. At this period the secret escaped, and in consequence thereof a patent was taken out in March, 1847, in the joint names of Mr. W. Millward and a person named Morris Lyons, who had acquired a knowledge of the secret." This application of bisulphide of carbon gave a considerable impetus to the manufacture of that unsavoury compound. In 1849 the use of selenide of carbon and a number of other agents for producing bright deposits was patented by a Birmingham plater named Stanhope Baynes Smith. Several other patents were taken out in the town between 1840 and 1849, for applications of electro-metallurgy and for improved methods of depositing various metals and alloys, by Alexander Parkes, O. W. Barratt, I. H. Russell, and J. S. Woolrich. They do not, however, need more than a passing reference.

## CHAPTER XXIII.

### SHIPS' SHEATHING AND BOLTS.

(*First Article.*)

COPPER SHEATHING — KEIR'S METAL — MATTHEW BOULTON AND THE ADMIRALTY — COLLINS' ALLOY— CHARLES WYATT'S PATENT FOR TINNED COPPER.

From a very early period naval architects have recognised the necessity of using some means of preserving ships' timbers from the attacks of marine animals, and for preventing, as far as possible, the growth of vegetation. It has been said that metal sheathing was known to the Romans, who are reported, on what seems good authority, to have used lead for that purpose. In the latter part of the seventeenth century sheet-lead was employed in the Royal Navy, one Thomas Hale having taken out a patent for rolling it. He wrote a pamphlet in proof of the valuable results which would accrue if the King's ships were sheathed with this invulnerable armour; and we believe that it had a trial of some considerable duration. The first vessel sheathed with copper was the Alarm frigate, built in 1761; but, according to Derrick's History, the practice did not become general throughout the Royal Navy until the year 1783. As we showed in our third article, Birmingham was very early in the field, and the alloy known as "Keir's Metal" was submitted to the Navy Board, in 1779, by the indefatigable Matthew Boulton. As the correspondence has never been printed before, we make no apology for reproducing it from the originals preserved amongst the Admiralty records. Boulton's letter, dated August 14, 1779, is as follows :—

"Mr. Boulton, of Birmingham, presents his most humble respects to the Earl of Sandwich, and begs his lordship's permission to wait on him, respecting the contents of the enclosed memorandum. He hopes that his lordship will be assured that nothing but the conviction of the utility of the proposal could have induced him to trespass on his lordship's leisure, or to add to the number of applications with which he is sensible his lordship must be daily troubled."

[*Memorandum.*]

"Matthew Boulton and James Keir have discovered a compound metal which they apprehend would be considerably better than copper for sheathing ships, and for making those bolts and nails and other pieces which are now made of copper. Their reasons are :—

1. That the metal, being much harder and stronger than copper, or any metal excepting forged iron, to which it is very similar in those properties, as also in its being sufficiently ductile and tough, is much fitter for the purpose of making bolts and nails, as they may be drove better and farther into the timbers, without being upset, as the carpenters call it; and when drove they are much less liable to bend, and consequently give greater strength to the ship.

"2. That the metal is less liable to be corroded or dissolved by salt water than copper is, and consequently would last longer.

"3. That they engage it shall not cost more than copper—perhaps less.

"4. That they request leave to have a trial made, and if their request be granted they will furnish a sufficient number of bolts, nails, and plates for that purpose."

The matter having been referred to the Navy Board, an experiment was ordered to be tried upon the Juno frigate, as appears from Keir's letter :

"London, October 8th, 1779.

"Honble Sirs,—In consequence of the permission which you were pleased to give that the bolts and plates made of a new metal invented by Mr. Boulton and myself should be tried upon the Juno frigate now building in Batson's yard, I have prepared a sufficient quantity of these bolts and plates for such trial, and shall be ready to attend at Batson's yard for that purpose on Monday next or any day after that which you shall be pleased to appoint.

"I am, &c., "JAMES KEIR.

"P.S.—I wait to receive the honour of your commands."

This letter is endorsed with a minute dated October 8, to the effect that "Mr. Keir was desired to attend the Board on Tuesday next, between ten and eleven," but unfortunately there is no further mention of the matter in the series of records from which the above was taken.

During the litigation in which Mr. Muntz became involved in order to protect his patent for sheathing, much was made of a patent granted in 1800 to William Collins for an alloy of very similar composition to Muntz's metal.

Collins was a Birmingham button maker, who, as he tells us in his evidence before a Parliamentary committee in 1799, subsequently migrated to London to engage in the copper trade. The earliest mention of his name amongst the Admiralty records occurs under date May 1, 1787, when he tenders for a supply of rudder braces and pintles of "pure wrought copper," similar to those which he supplied in 1785. The price was to be 14½d. per lb., and amongst other advantages it was stated that their value would be 2½d. per lb. more than that of "mixed metal." This would appear to show that some alloy of copper was even then used for ships' fittings, though, of course, it may have been nothing more than "gun metal," which has been known for centuries. In 1790 we find that the Board gave an order to copper the bottom of the "Prosperity" with sheets of different sorts, which were to be inspected every six months. Several of the reports are extant, but unfortunately no particulars are given of the composition of the various sheets. At the end of four years "the sheets of the planished or improved sort were very clean, and the surface smooth; those marked No. 6, not planished, were in general clean, but not so much as the former. The other, a *brass colour*, appeared the most foul and corroded, and the nail heads in the copper were particularly foul with mussels and long weeds." Collins's patent was not taken out until 1800, but it would seem from the above that he was experimenting upon alloys suitable for use as sheathing for some time before. A good deal of correspondence follows, and Collins was anxious to see the sheathing in the condition it was in when taken off the ship, but it appears to have been fired to remove the weeds, &c., before he had an opportunity of doing so. He endeavoured to persuade the Board that his metal would be found "a valuable acquisition to the sheathing service of his Majesty's navy," but on the whole it was not a success, and the trials were finally abandoned in 1800. By a minute of the Board, dated March 10 of that year, we learn that the "improved copper being very hard rolled it is found so stubborn as not to set so close to the bottom as the more malleable kind, and on the bottoms of several ships lately taken into dock it has been found so very much corroded as to make it absolutely necessary to change it, although it had not been on more than three or four years. We therefore are clearly of opinion that so far from it having a preference to the copper sheets formerly in use, that from the shortness of its duration and stubbornness, it is really inferior, and as we have paid an advanced price of £2. 6s. 8d. per ton for it, we propose as soon as the stock now in hand is expended to return to the use of the common sort." On the back of this document there is a column of figures which seem to refer to the quantities of sheathing supplied by Collins. The total value is £3,225.

In 1790 a Birmingham merchant named Charles Wyatt, in conjunction with the William Collins just mentioned, took out a patent for tinning copper for ships' sheathing and other purposes. A considerable trade appears to have been carried on in this material, as we showed in the article which appeared on the 17th of June. In the year 1796 Wyatt submitted his tinned copper to the Navy Board in the following letter:—

"New Bridge St., June 17, 1796.

"Honourable Sirs,—Being persuaded, from my own experiments, that the tinned copper sheets which I manufacture would be much more durable for sheathing vessels than common copper; but being aware at the same time that the truth of this opinion may be justly questioned on the score of interest, I take the liberty of presenting to you the means of ascertaining the fact from your own observation.

"This liberty I take, on the presumption of its being greatly to the interest of his Majesty's Service to decide, whether there is or is not a better material for sheathing than the copper now made use of.

"The two vials herewith contain sea water. If you will be pleased to put into one vial the piece of red copper, and into the other the piece of tinned copper, and notice the change that takes place in the water, it will afford a certain proof of a correspondent change in the two metals.

"The moment the water becomes blue, the copper begins to dissolve, and consequently, to decay. That vial, therefore, which exhibits no marks, or the fewest marks of this decay, must necessarily contain the most lasting material.

"But, although this experiment will clearly show the superior durability of the tinned copper, it may still be urged that it will grow foul at sea.

"To this I should reply that there is no fact whatever to support such an assertion, and that if every imaginary objection to improvement is first to be removed improvement would become impossible.

"The objections to the present mode of sheathing were, at first, much stronger than any that can be urged against that which I wish to establish.

"The tinned copper, however, from its resisting the action of salt water, being undeniably more durable than the other, it follows of course that less strength of sheathing than what is now used, would answer the same purpose, and therefore that the present expense may be diminished; and I conceive that 10 or 15 per cent. may be saved by using the tinned instead of untinned copper.

"Another advantage, and perhaps not one of the least, that would result from the use of the tinned sheets would be that the tinning on the side next the vessel, by preventing the corrosion of the copper, would preserve any copper or iron work that might come into contact with it.

"On duly examining this subject, therefore, I trust that you will think the new material deserving of a fair trial on some of his Majesty's ships. There is certainly a probability of advantage, and none, that I can foresee, of a disadvantage.

"I am, &c., "CHAS. WYATT."

It appears from a minute on the back of the document that the inventor was desired to attend the Board "To-morrow at 2 o'clock," but there is no record of the result of the interview, nor does it appear that the tinned sheets were ever submitted to trial. The address, it should be stated, was that of Wyatt's London office.

## CHAPTER XXIV.

### SHIPS' SHEATHING AND BOLTS (CONCLUDED).

MATHEW BOULTON AND THE COPPER MONOPOLY — ROLLING SHEATHING FOR SHIPS — HIS COPPER TOO PURE — SUGGESTED EXPERIMENTS — INTERESTING CORRESPONDENCE — KEIR'S METAL UNSUCCESSFUL — MUNTZ'S METAL—COSTLY LITIGATION—METAL BOLTS FOR WOODEN TRENAILS — COLLINS AND WESTWOOD PATENTS—THOMASSON'S COPPER-COVERED IRON BOLTS —MUNTZ'S METAL BOLTS.

Matthew Boulton's efforts to break down the copper monopoly which was seriously threatening the trade of the town at the latter end of the last century are well-known. So far as we are aware it has nowhere been recorded that he subsequently turned his attention to the rolling of copper for the purpose of sheathing ships. In the year 1800, when he was 72 years of age, he started a new manufacture at Soho, as appears from the following correspondence with the Navy Board, now printed for the first time :—

"Soho, near Birmingham, Sept. 6, 1800.

"Honourable Sirs,—In reply to your favour of the 5th inst., the four sheets of copper you have received were intended to have been delivered by me, but I unfortunately had left town before their arrival. In consequence of the conversation that passed at your honourable Board I ordered four sheets to be rolled from my refined copper as specimens of the proposed sheathing.

| | lb. | oz. |
|---|---|---|
| A.—Are two sheets, the one is *soft* the other hard rolled to the thickness of 28 ounces per square foot, to the breadth of 14 inches, and length 4 feet ........................ | 16 | 6 |
| B.—Are two sheets, the one is rolled soft and the other hard to the breadth of 14in., length of 4 feet, and 32 ounces per square foot ........................ | 18 | 11½ |

"I should be happy to receive your remarks respecting these sheets, and if you please to honour me with an order for as much as will sheath one ship it shall be as smooth rolled as any you have used, and it shall be rolled to any degree of hardness or softness that you may think proper, and I have not the least doubt but I can refine it so as to make it as durable as the best copper sheathing that has ever been applied to his Majesty's ships.

"Waiting your further commands.

"I am, &c-, "MATTW. BOULTON."

"Soho, Nov. 20, 1800.

"Honourable Sirs,—Sensibly impressed with the favour you did me in giving me an order for sheathing, I find no small difficulty in replying to your letter of ye 18th inst., but I trust upon a simple statement of facts you will pardon my delay.

"Upon the receipt of your letter of ye 8th September I immediately put your order for sheathing in hand, and before it was compleated the East India Co. gave notice that they wished to receive tenders for some hundred tons of sheathing, which (as my stock of copper was large) I could not resist so favourable an opportunity of disposing of it. I was but too successful in my application, and I obtained the whole contract; my only difficulty then is, that in the execution of *their order* I am bound by my contract to deliver on shipboard a considerable part by the end of the present month; which I could not have accomplished had I not appropriated that very copper intended for *you*, to the fulfilment of the first part of my contract with *them*: I will, therefore, thank your honours to allow me a little more time, and to inform me the latest period you can indulge me with; when you may depend upon all my exertions that you may not be disappointed. This being my first experiment in the manufacturing of sheathing, I could not form an accurate opinion of my powers of execution previous to this essay; but finding I cannot execute it with so much expedition as I had hoped to do, I am determind to erect in the course of the ensuing summer a better apparatus than has yet been in use for the purpose; and I have full confidence that my powers then will be equal to my most sanguine wishes, and I shall then hope that you will allow me an opportunity of retrieving my charactor, which has in this instance failed, of punctuality.

"I am, &c., "MATTW. BOULTON."

The Board were rather annoyed at this diversion of the sheathing intended for the service of the Government, as is evident by the tone of the minute of their reply to Boulton :—

"Nov. 24, 1800.

"Acquaint Mr. Boulton that we were desirous of making an early trial on the copper manufactured by him and ordered for export the quantity named in our letter of 8th Sept. that a fair and partial trial might be made, and an early report received—but as he has diverted the copper appropriated for this trial into another channel, he has lost the opportunity we intended for his advantage, and as the ships for which it was intended are now actually ready for coppering, any later period for them cannot be granted; but if he continues desirous that an experiment on his copper should be made in the King's service we shall have a couple of frigates ready by the end of January, and desire he will send up the quantity already demanded of him to Deptford Yard by the 30th of January at the latest."

On the 8th of January he sends up 3,300 sheets, weighing about 12 tons, to Deptford, and orders were given to use them for sheathing one side of the Æolus frigate, the other side to be covered with common copper. The trial was unfavourable, and Captain Gore, of the Medusa, the other frigate mentioned in the minute, reports in 1802 that Boulton's copper was extremely foul, with long grass adhering to it in several places, whilst that on the other side of the vessel was perfectly clean. This was communicated to Boulton, who writes under date December 24, 1802 :—

"Honble. Sirs,—I beg your pardon for having delay'd answering the honor of your favor of the 26th ult., which has proceeded from a long and severe indisposition ; having been confined to my bed chamber from the beginning of March to this day.

"I am much disappointed in my expectations respect'g the copper sheathing which you say was applied to one side of his Majesty's ship Medusa, I have always been of opinion that the durability of copper sheathing depended on the purity of that metal, and I assure you the whole quantity I had the honor to furnish you with was doubly refin'd, for it was made from the scissel, or scrap copper, remaining after cutting out the halfpence I made by order of our Government. Such coin can only be made from fine tough cake copper that will roll both hot and cold, and is capable of being afterwards coined. This scissel, or scrap copper, being remelted and refin'd a second time, has always been found to be more pure than the tough cake from which it was originally made. But as all theories and reasonings are liable to prove fallacious when put in competition with actual experiment, I think it wou'd be well if your Honorable Board were to order a regular series of experiments to be made on this subject, as it is not only desirable, but most highly important, to ascertain what qualities in copper make it bad, and what make it good, for the purpose of sheathing—such as

Experiment 1st.—To try copper as highly refin'd as it can be, hot rolled only.

2nd.—Ditto ditto, finished rolling cold in hard polish'd rollers.

3rd.—Fine copper alloy'd with 1 or 2 per cent. of lead, roll'd hot only.

4th.—Ditto ditto ditto, rolled cold, as before.

5th.—Query what are the effects of painting the sheathing with white lead and oil on the side that goes next the vessel?

It is said that the poisonous solution of copper prevents barnacles from adhering to the sheathing, as well as all vegetable matter from growing thereon; if so, perhaps some coarse species of copper may do better than the finest, but this reasoning does not accord with my ideas. If copper be alloy'd with iron or zinc, I know such alloys will accelerate its destruction. If alloy'd with tin it becomes brittle, and will neither roll hot nor cold. My copper sheathing was all roll'd cold the last time of rolling it in order to make it smooth; but it was afterwards anneal'd soft, in order to render it the more pliable in its application to the ship's sides; but I am of opinion that it would be less liable to corrosion if it was not anneal'd after the cold rolling. I am inclined to think that the accelerated corrosion of iron bolts in the vicinity of copper will be better understood when we become more acquainted with the galvanic fluid, which I am persuaded is a principal agent in that operation.

"When the Medusa returns I shou'd wish to see a sheet of her copper with my mark on it taken off, as well as a sheet from the opposite side of the vessel, as something may present itself to our observation from thence capable of throwing some light on the subject, for, although I profess to understand metallurgy, I have very little knowledge or experience of what relates to shipping. But I should be happy to obey your commands on all occasions, or to assist in making any experiments, and remain, &c.,

"Mattw. Boulton."

Boulton was informed that when the Medusa returned to England he should be supplied with sheets for the purpose of examination; but this was never done, as the copper was removed from both ships and melted down, the officers of the yard pleading that they were unaware that the ship had been coppered for experiment.

These letters are wonderfully characteristic of the man. The latter was written when he was in his seventy-fifth year, and just after his recovery from an agonising attack of his old enemy—stone in the bladder. Notwithstanding this there is an amount of vigour and freshness about them which rarely belongs to an age when a man's strength is beginning to be but "labour and sorrow." There is one passage which seems to show that the experiments with

Keir's metal were unsuccessful, for he says "if copper be alloyed with zinc I know such alloys will accelerate its destruction." Keir's composition, it may be remembered, consisted of copper, zinc, and iron. His remark that the corrosion of iron in the vicinity of copper would "be better understood when we become more acquainted with the galvanic fluid, which I am persuaded is a principal agent in that operation," was profoundly scientific, for it must be remembered that Voltaic action was then but little understood.

This article would be incomplete without some reference to Mr. Muntz's sheathing, a patent for which was granted in October, 1832. It was of enormous importance to the shipbuilder, besides being a source of great wealth to the inventor, who realised £70,000. profit, as stated in his petition for a prolongation, which, by the way, was refused. The patent was attacked several times, no less than seven bills having been filed, six actions commenced, and a writ of *scire facias* obtained to repeal it. Mr. Muntz was uniformly successful, and his opponents had to pay £9,000. in costs and damages, but his own costs amounted to £8,400. Many eminent authorities were of opinion that the alloy was not "a new manufacture," and was therefore not a proper subject for a patent, a metal of the same composition having been described in Collins's patent of 1800, besides being mentioned in several well-known scientific works. This part of the question was fully dealt with in our third article. Curiously enough, Messrs. Grenfell, who subsequently took the matter up, wrote in 1837 that "the patent was all humbug!" Some very "nice points" —so dear to the hearts of the lawyers—arose during the litigation, which is, however, very instructive even to ordinary persons who are interested in manufactures. It has been made much of by the opponents of "preliminary examination" of patents, who urge that it would never have passed such an ordeal. This is no doubt perfectly true, and yet the patent was held to be a valid patent by some of the most eminent judicial authorities of the day. A very full account of the litigation may be found in Webster's "Reports of Patent Cases," and the matter was discussed at considerable length by Mr. Aitken in his contribution to "Birmingham and the Midland Hardware District." Some details of the early difficulties of the invention are given in two articles in the *Mechanics' Magazine* for October 12 and November 2, 1839.

The use of metal bolts as a substitute for the old wood trenails for fastening ships' timbers together was attended with considerable danger. The wood was liable to become what is known as "iron-sick," that is, the portions adjacent to the bolt decayed, allowing the fastenings to get loose. Shipbuilders then bethought themselves of copper bolts, but it was long before they could be obtained of sufficient toughness to allow of being driven. This was at last accomplished, "after great pains and labour, by two ingenious artists of Birmingham, who found out the method of making copper bolts far superior to the very best iron ones ever made," to quote the evidence given by Mr. Thomas Williams, M.P., before a Parliamentary committee, in 1799. "The Navy Board," he continues, "after giving them the fullest trials, approved and adopted them, and they have ever since been universally used, to the incalculable advantage of the British Navy." No names are given, but we are pleased to be in a position to identify the "two ingenious artists of Birmingham." One was the William Collins mentioned in the previous article, and the other John Westwood, a button-maker of the town. Collins's plan, which was patented October 2, 1783, consisted in forcing a round rod through a diminishing series of holes, the necessary force being obtained by grooved rolls, roughened so as to give a firm grip. The bar was then cut into lengths, each sufficient to form a bolt. There is a fundamental error in another part of the specification, for the inventor proposes to make bolts by covering iron with copper or other metals. It was not then known that the juxta-position of iron and copper in sea water was to place the copper in the most unfavourable condition for resisting corrosion. Westwood's patent was taken out about six weeks after Collins's, and the claim broadly stated was for making metal rods by the use of grooved rolls, the shape of the grooves varying according to the section of the rod required. The principle was not new, a patent for the same thing having been granted to John Purnell, of Frampton-on-Severn, Gloucestershire, in the year 1766.

Before taking leave of this subject it should be mentioned that Sir Edward Thomasson thought he had hit upon a grand idea when he commenced to manufacture ships' bolts covered with copper by drawing a tube of the protecting metal over an iron rod. He says, in his amusing *Memoirs during Half a Century*, "the shipbuilder was in ecstasy when he discovered that the bolts drove in well without stripping, but we had both forgotten the action of the muriatic acid of the sea water insinuating itself, by degrees, between the iron and copper, which in a short time would render the bolt slack and unsteady, which prevented a second trial."

The application of Muntz's metal to bolts formed the subject of a separate patent, granted in 1832, a few months after the original sheathing patent. There was a long discussion in the *Artizan* about the year 1855, on the disadvantage of fastenings of that metal, which it was alleged was liable to lose its fibre and to become crystalline and brittle. Certain sorts of brass, as is well known, are liable to such a change of structure and consequent deterioration, under particular conditions.

## CHAPTER XXV.

## UMBRELLAS AND PARASOLS.

UMBRELLAS IN 1780—UMBRELLAS AS FANS—FANCY UMBRELLAS—PARATOUTS, PARASOLS, PARAPLUIES—UMBRELLA FIRE SCREENS—THOMASON'S RHABDOSKIDOPHOROS—LADY THOMASON'S "AD LIBITUM" UMBRELLA—TUBULAR STICK UMBRELLAS—HOLLOW TWIG UMBRELLAS—RUBERY'S PATENTS—HOLLAND'S HOLLOW-RIB PATENTS—FOX'S PARAGON RIBS.

IT is difficult to write about umbrellas in a serious tone, but in view of the lofty aim of these articles we are bound to be historical and perhaps dull. The moral and social aspect of the umbrella has already been dealt with many times, and an amusing little book on the subject was issued some years ago by Messrs. Sangster, the well-known manufacturers of the article. The general history of the umbrella is well treated in the introduction to the "Abridgments of Patents relating to Umbrellas and Parasols," published by the Commissioners of Patents. All we have to do is to deal with the local part of the question. The manufacture of umbrellas, or rather of umbrella "furniture," that is, the divers parts of the skeleton, which are afterwards put together by the maker, was commenced in the town at a very early period, probably about the year 1790. The first mention of umbrella furniture, as a separate branch of business, is, as far as we are aware, contained in Bisset's *Magnificent Directory* of the town, published in 1800. We there find the names of John Boyce, umbrella maker and brassfounder, Deritend; and William Cox, upholder, cabinet maker, and umbrella maker, Bordesley. In 1808 there were five firms engaged in the business, according to the same author's *Magnificent Guide*, the additional names given being those of John

Turner and Co., Temple Street; Rubery and Barrs, and James Boyce of Alcester Street. The last-mentioned firm is still in existence, and until within a few years there were two houses of the name, which is one of the oldest in the town in connection with brassfoundry, the firm of Boyce and Radcliffe having been in existence 110 years ago. Their place of business was in Freeman Street. The name of Rubery also is well-known in Birmingham in connection with improvements in umbrellas, as we shall subsequently show. Neither umbrella furniture or umbrellas are mentioned in Sketchley and Adams' *Directory* of the town, issued in 1770, but that is in itself no proof that the manufacture was then unknown here. The practice of carrying umbrellas, which is said to have been commenced, so far as the male sex is concerned, by Jonas Hanway, about the year 1750, was no doubt of slow growth. They had, however, been used by women long before, and Gay, writing in 1715, refers to them in that way. As soon, however, as any large demand arose for the article, it is pretty certain that makers would turn their eyes to Birmingham as the place of all others where the numerous small parts which go to make up an umbrella could be manufactured with economy and advantage. It will be seen, from what has gone before, that the umbrella was not an invention so much as a fashion, but one which bids fair to be permanent.

The first patent in which umbrellas are mentioned is that granted in 1780 to Mark Bull, saddler, of Mark Lane, London, for a support to be attached to a saddle or carriage for the purpose of holding an umbrella. Two patents were issued in 1786 to a London "umbrella maker," and a Bristol clock maker, respectively. The next is dated 1802, the patentee being Abner Cowell Lea, "manufacturer, of the parish of Aston," the object being the substitution of "rods or bars of iron or steel in various parts of the construction of umbrellas, parasols, or other instruments or utensils of the like nature, instead of brass or other less strong, light, and durable metals heretofore used." Sheet metal is first cut into strips by a pair of circular shears, and then drawn through dies to the shape required, the notches in the stretchers being formed by circular saws or cutters fixed in a lathe. A sheet of drawings is annexed to the specification, and the various processes are described with such an amount of detail and in such a way as to show that the trade was then of considerable extent and importance.

In the same year a patent was granted to John and Joseph Barnett, the former being a toy maker of this town, and the latter a cutler, residing in Warwick. The invention is described as "an improved mode of making parasols and umbrellas, whereby the same may be used for fans and for other purposes." The inventors claim the privilege of calling it "Barnett's patent paratout," the distinguishing peculiarity being that some of the ribs may be released whilst the others are relaxed, which allows the apparatus "to assume various useful and fanciful forms."

The use of this word "paratout"—meaning that which protects us against anything and everything—sun, rain, snow, hail, and wind, reminds us to note that "umbrella" and "parasol" have in reality the same original meaning. They both indicate a shade or protection from the sun, though why the former term should have been appropriated to a shield from the rain, whilst a special word should have been introduced to denote a "sun-shade" is more than we can say. The French has *parapluie* and *parasol*, which are formed on the model of many similar words, such, for instance, as *paravent*, *paratonnerre*, meaning something which protects from the rain, sun, wind, lightning, &c. Oddly enough, *ombrelle* means a parasol, thus presenting a pitfall for the unwary translator who renders literally. The Germans, with characteristic independence, speak of their *Sonnenschirm* and *Regenschirm* (sun or rain screen), or sometimes *Schirm* alone. Our word "sunshade," which was introduced, if we remember rightly, about twenty years ago, presents a similar instance of what may be called a home-made word. To return, however, to the paratout, which is a very Proteus, and is shown in the form of a canopy, the front ribs being extended and those behind hanging down, thus forming a capital protection to the back when the rain comes from behind. When a fan is required the umbrella is flattened out, as it were, the lower portion of the stick being detached by a suitable arrangement, thus giving a shorter handle, and offering greater facility in wielding it. The umbrella may also be transformed into a fire or candle screen. As may readily be understood the mechanism is very complicated, and it must have been very heavy to carry and particularly liable to get out of order.

The next on the list is the "rhabdoskidophoros," or walking-stick umbrella, of the ingenious Sir Edward Thomason, who patented the invention in 1808. As its name imports it is nothing more than an umbrella capable of conversion into a walking stick, there being no improvement in the mechanism. For this purpose the "umbrageous part" is slid into a hollow tube of metal or any other suitable material. The contrivance was clumsy, and since that time we have been able to make an umbrella which serves the purpose of a walking stick by simply rolling them up and securing them by a band. The original umbrellas we may here mention were not intended to walk with, and generally had a large ring, where the ferrule is now placed, for the purpose of hanging them up when not in use.

Just about this period there seems to have been a mania for contrivances of this sort, a patent having been granted to McGregor and McFarlane, of London, about two months after Thomason's patent, for the very same thing. The fact of the phrase "umbrageous part" occurring in the London patent would seem to show that the inventors were not working in ignorance of what was being done in Birmingham.

The "ad libitum" umbrella, patented in 1809 by Phillis Bown Thomason, wife of the aforesaid, was a much better looking article. The object was the same, but the covering was telescopic and made to slide entirely off the umbrella, whereas in the former patent the sliding tube formed a part of the handle. The wording of the specification is identical in places with that of the husband's patent, and it is probable that Thomason had as much to do with it as his wife. The idea was suggested no doubt by the telescopic hearth brush and toasting fork of which he was the inventor, and which he patented in 1803. Lady Thomason, we may here mention, lived until a good old age, and died in July, 1851, at Malvern, in her 80th year.

Robert Davis, who describes himself as an "umbrella and parasol furniture maker" of this town, took out a patent in 1811 for making the various parts of umbrellas of "rolled gilding metal" instead of "dipping metal," and he describes various modes of preparing the metal and colouring the furniture by heating and "pickling" it, the details of which are of purely technical interest. Samuel Hobday's patent of 1821 seems to indicate that the umbrella was receiving attention from mechanicians, the early examples being confessedly very imperfect, especially in the mode of connecting the ribs to the "top notch," and the stretchers to the "runner" or sliding piece which moves up and down the stick. It is the object of this patent to improve the method previously in use. The inventor was a maker of snuffers, for which he took out two patents, in 1810 and 1821 respectively. He belonged to a family which has long been connected with the manufactures of the town. In 1823, Francis Deakin, wiredrawer, patented the use of hollow metal sticks covered with papier mâché, varnished or japanned to prevent oxidation of the iron or other metal employed. The other parts of the umbrella were also to be protected by similar means. The various members of the Deakin family have taken out numerous patents from time to time, and there appear to have been two different Deakins of the same Christian name who were contemporary. J. G. Hancock, "plated beading and canister-hinge manufacturer," patented, in 1826, a mode of making sticks by boring out the pith from the centre of an osier twig, and supplying its place with an iron wire or hollow tube. In this manner a very strong yet light stick was obtained. The inventor must surely have been related to the celebrated diesinker of the same name who also appears in the patent records several times.

Reference was made in an earlier part of this article to the fact that the name of Rubery was well known in connection with the umbrella trade. Patents were granted to Mr. J. Jeremiah Rubery, of this town, in the years 1836, 1837, 1842, and 1844, and to Mr. John Rubery in 1853, 1857, and 1858. These consist for the most part of improved methods of making the various parts of umbrellas by the use of stamps of various kinds. The later patents include claims for making certain parts of umbrella furniture of zinc, and for covering such articles with brass by the agency of voltaic electricity. The inventions of Mr. J. J. Rubery display great ingenuity, and his patent of 1837 furnishes an interesting instance of the commercial value of a method of producing an exceedingly trivial article. The patent was for making small ferules called "top tips" for placing upon the ends of the ribs of the old whalebone umbrellas to prevent them from splitting when drilled for the wire which connected them to the "top notch." The specification was found to be slightly defective, and it was accordingly amended by a "memorandum of alteration" shortly after the grant was made. Subsequently an action was brought by the patentee for infringement, when the jury decided that certain parts were not new. These were disclaimed in 1838, and in 1841 an order was obtained from the Master of the Rolls to correct a clerical error in the enrolment. Including the original expense of the patent these proceedings must have cost some hundreds of pounds, which large expenditure was incurred about a minute detail, the existence of which hardly one possessor of an umbrella in a thousand would be aware of.

A great step was the introduction of hollow instead of solid ribs, which was due to Mr. Henry Holland, of Darwin Street, who patented the invention in the year 1840. Ten years later he obtained a patent in which hollow ribs and stretchers of an elliptical section were described. Mr. Holland's inventions also comprise certain modifications of the ordinary modes of connecting the ribs to the stretchers, which are rendered necessary by the substitution of tubes for solid rods. These umbrellas were shown at the Exhibition of 1851, and in the Report of the French Jury they were referred to as being the only English improvement worthy of being noticed. With that jealousy which so frequently manifests itself when "perfidious Albion" stands in the way, the report went on to observe that the invention was not new, a patent for a similar thing having been obtained by one Pierre Duchamp, of Lyons—a considerable seat of the umbrella trade in France—in the year 1846. By a "certificate of addition," dated July, 27, 1847, the inventor claimed trough-shaped ribs. The reporter had not taken the trouble to ascertain that Holland's patent was anterior to that of the French claimant. Ingenious and valuable as Mr. Holland's frame was, it was improved upon by Mr. Fox, of Deepcar, near Sheffield, who patented his "paragon" umbrella, in 1852. It consists, as is well known, in making the ribs and stretchers of a trough-formed section, thus securing great strength, stiffness, and lightness. It was a beautiful invention.

Here we bring our notice of umbrellas to a close, although we have left many inventions unnoticed; but the subject is not one which lends itself to anything like popular treatment. Many of the improvements which have been made in "the Englishman's constant companion"—as it has been called by those who live in countries where the weather is more settled than it is here—are too minute to be appreciated by any but experts.

## CHAPTER XXVI.

### THE STAMP AND PRESS.

The Press and Buttons—The Stamp and Watch Plates — Stamped Chasing — Ford's Patent — Stamped Brassfoundry in 1777—The Fly Press and Hancock's Patents — Presses by Power—Stamps by Power—Large Stampings—Griffiths's Patent — Hetherington's Patent —Hill's Pneumatic Hammer—Fleetwood's Patent.

It may, we think, be taken for granted, that those invaluable tools, the stamp and press, were first used for fashioning coins, which may be said to form the earliest examples of "manufactures in metal," using the word "manufacture" as denoting a large number of articles precisely identical. They are said to be of French invention, and they were introduced into the English Mint in 1561, but they did not come into exclusive use until a century afterwards. They were employed here early in the last century, in the button trade—the press for cutting out the "blanks" from the sheets, and the stamp for impressing with a pattern or device; but when they were applied generally to the production of a number of similar articles we are unable to say. So far as we are aware, the first patent for the application of the process other than to the manufacture of buttons, is that granted to one George Sanderson, an Exeter watchmaker, in the year 1761. This patent is interesting for two reasons: first, because it is a very early suggestion for the manufacture of watches by machinery, and secondly, on account of the locality—Exeter and the West of England generally not being in the least distinguished for mechanical genius. There is, however, no hint of the application of stamping to any other purpose than making parts of watches—the shoemaker, in this case, not going beyond his last. The patent which probably fixes the date of the stamped brassfoundry trade is that granted to John Pickering, a London gilt toy maker, or jeweller, as he would be called in common parlance, in the year 1769. He had a perfectly clear idea of what he was about, as is evident from the title of his patent which—as was usual in those days—contains an epitome of the invention. It runs: "A new method of performing that kind of work commonly called chasing, for the working in gold, silver, brass, tin, and other metals, of various things, particularly coffin furniture, ornaments for coaches, chariots, sedans, and other carriages; for cabinet work and other domestic furniture, and other articles, by a machine consisting of an oblong square frame, with two rods, in which a moving forcer is worked upon a striking block, with a die fixed thereon, formed for each respective purpose, whereby the work is executed in a much more expeditious manner, and far superior in beauty and elegance to anything of the kind (not being actual chasing) ever yet performed by any other means."

In the same year (1769) a patent was granted to Richard Ford, of this town, for rolling and stamping metals, the latter part of the invention consisting of "raising by a stamp and press scale-pans, saucepans, warming-pans, basins, plate-covers, kettles, ladles, and various other things, out of silver, copper, and other metals." The description is very meagre, the inventor only saying that he proposes "to raise, by a machine generally called a stamp or press, scale-pans and the aforesaid and sundry other things, by two implements called dyes, placed under the hammer of a stamp or screw of a press, the one being concave and the other convex, which, by the pressure of the hammer or screw forced into the dye, the shape or form of the thing designed is accomplished." Although the inventor is simply described as a "smith," we believe that he was identical with the person of the same name who subsequently built that sham Gothic structure, Hockley Abbey, and was well known in the town as a very ingenious and prosperous manufacturer. He invented a single-wheeled chaise—the idea was in great favour at the time—for which the Society of Arts gave him a gold medal. The "Abbey' is mentioned in Charles Pye's description of Birmingham, and also in Bisset's *Poetical Survey*. But the old *Directory*, which has been so frequently referred to in the course of these articles, seems to say that there were two Richard Fords in Birmingham at that time. We may here mention that an elaborate patent for stamping hollow-ware was granted to John Browne, of Llannedy Forest, Carmarthen, in 1776.

An interesting patent was taken out in 1777 by John Marston, brassfounder, of 2, Paradise Row (now "Street"), and Samuel Bellamy, engraver and diesinker, also of this town. It perhaps marks the date of the introduction of the stamped brassfoundry trade into Birmingham, although there was little originality about it when compared with Pickering's patent, just referred to. The title is, "A method of stamping upon plated metal, gilt and other metals, all sorts of hat and cloak pins, and all kinds of figures, decorations, ornaments, and other devices for cabinet furniture and lock furniture, and intermixing divers colours in the same." The stamped shells were to be filled with soft metal alloy, the edges being turned over. A shank, nail, or screw, serving to attach the ornament, was to be embedded in the filling, or attached in some other manner. The colours referred to in the title are produced by perforating the ornaments and placing foil at the back, so as to show through. The goods when finished are to be lacquered. The word "brass" does not occur anywhere in the specification, although it would, of course, be included in the expression "other metals." Mr. Aitken, who refers to this patent in the chapter on the brass trade, contributed to "Birmingham and the Midland Hardware District," thinks this was done in order to steer clear of Pickering's patent, but that can hardly have been the reason, if, indeed, the omission was intentional. At this date the trade was in all probability well-established, and the patent of James

Yates, of Bordesley, brassfounder, taken out in 1788, must be regarded as being very much behind the age. The title is "a new method of multiplying engravings or chasings on all kinds of metals, particularly applicable to the engravings or chasings on the ornaments of coaches and coach harness, and many other purposes where engravings or chasings are required." The description discloses no novel features whatever, the process being that of Marston and Bellamy, with a few additional details. This Yates we take to be the same person who subsequently became a member of the firm of Yates and Hamper, a leading house in the brassfoundry trade for many years. The name is still connected with the brassfoundry business, though perhaps by a mere accidental coincidence.

Passing over the patents for the simple application of the process of stamping to the formation of certain articles, we have to notice J. G. Hancock's patent of 1804 for working a fly press. He dispenses altogether with a screw, and substitutes for it what may be called a "face cam," which presses down the bolt of the press as it revolves, the bolt being brought up again by the action of a spring. By this mechanism the press may be driven by the continuous rotation of a shaft in the same direction, there being obviously no necessity to reverse the motion of the cam. Strictly speaking the contrivance is little more than a "mechanical equivalent" for the ordinary screw. In 1809, Samuel Whitfield, brazier and scalebeam maker, of Church Street, had a patent for stamping ears and handles for culinary utensils, which presents some features perhaps novel at that time. The enormous expenditure of muscular force necessary to work heavy stamps attracted attention at an early period, and Edward Biggs, a brassfounder of the town, patented a mode of working them by power in the year 1813. The prime mover drew up the stamp to the top of the framework, and held it here until released by a cord placed near the hand of the workman. When the blow had been delivered, a self-acting mechanism threw the winding arrangement into gear again, and the stamp was raised in readiness for a second blow. Biggs subsequently patented a method of stamping various articles, such as warming pans, frying pans, and the handles for the same. We may also mention the patent of Henry Pershouse, taken out in 1818, for stamping scale pans.

As the art continued to be improved, manufacturers were enabled to produce larger articles, and to raise them still more, but there was a difficulty with regard to the uniform distribution of the material. When it is remembered that a dish-cover, for instance, is nothing but a flat metal plate, to begin with, it will be seen that the material must undergo very severe stretching before it arrives at its final form. Further, it may readily be supposed that this stretching does not take place uniformly, so that some places would be thinner than others. These remarks will be thoroughly understood by those who have watched the operation of making a "raised" pie out of a flat sheet of dough. The parallel is not quite perfect, but sufficiently so for our purpose. The difficulties above alluded to were cleverly overcome by Mr. T. F. Griffiths, who took out a patent for his improvements in 1846, and he succeeded in obtaining vessels of equal thickness throughout. This was accomplished by a peculiarity in the dies, which were so shaped as to prevent the "raising" from being carried out solely at the expense of the sides, which were thinned to an inconvenient extent by the old process. Mr. Griffiths was also the patentee of a method of making tea-kettles in one piece by a very ingenious combination of the processes of stamping and "spinning," as it is called.

Another noteworthy improvement was that originated by Mr. William Hetherington, in the year 1852, the essential feature of which consisted in the pressure by an annular die of the circumference of the plate to be operated upon in such a way that it was gradually drawn from between the pressing surfaces, as it was forced into the lower die by the pressure of the top or descending die. In this manner the formation of puckers or creases in the metal was prevented. The force of the blow was increased by the pressure of steam actuating a piston in a cylinder analogous to the arrangement of a steam hammer. This reminds us that Mr. Joseph Hill obtained a patent for a pneumatic hammer to be used for stamping metals in the year 1852.

In 1859 Mr. D. J. Fleetwood patented a method of producing, by pressure, vessels having more or less the shape of a sphere or spheroid, or with a spheroidal body and taper neck, or prolongation. He takes a hollow seamless cylinder of metal closed at one end, and fills it with soft metal. The cylinder is then subjected to pressure in suitable dies, and the soft filling metal escapes through the open aperture. He is also the inventor of a combined hydraulic stamp and steam hammer, for producing large articles, such as dish-covers or basins, which it would take us too long to describe intelligibly. These machines were patented in 1859 and 1866.

## CHAPTER XXVII.

### MISCELLANEOUS INVENTIONS, 1759-1780.

BLOCKLEY'S WHEEL TYERS—SAMPSON AND FREETH'S STEEL MILLS—WHATELEY'S PATENT WIRE—OPPENHEIM'S RED GLASS—WHITEHURST'S SECRET HINGES—DEARMAN'S "BRUMMAGEM" HOES AND MALLEABLE CASTINGS—TAYLOR'S CAST IRON POTS—TALBOT'S SCRAPING MACHINE—LINDOPP'S PORTMANTEAU SADDLE—BACHE'S PERPETUAL PROPELLER.

HAVING dealt in a consecutive manner with several of the staple trades of the town, it now remains to gather up the fragments by briefly describing the miscellaneous inventions which have originated here, and which have found no place in previous articles.

We commence with the patent granted in 1759 to Thomas Blockley, who is described as a locksmith, for rolling iron into different forms and making wheel tyres. In the "Directory" for 1770, the name appears under "locksmiths," the address being 61, Bull Street. It would appear from the patent taken out in 1768 by Sampson and Samuel Freeth, merchants, for grinding mills, that the manufacture of "steel mills" was well established here more than a century ago. The patentees carried on the business, in combination with that of edge-tool makers and factors, at No. 69, New Street, which was then of quite a different character to what it is now. The drawing attached to the specification shows an ordinary hand mill, suitable for being attached to a post; the grinding mechanism consisting of a case-hardened roller, made slightly taper and covered with teeth, increasing in fineness from the smaller to the larger end. The box in which the roller turns is also provided with similar teeth, the fineness to which the meal is to be ground being regulated by a set screw which moves the roller in the direction of its axis, and thus increases or decreases the distance between the grinding surfaces, the box being also tapered to correspond with the roller. This mode of adjustment is similar to that adopted by Archibald Kenrick, of West Bromwich, in his well-known coffee-mill patented in 1815. With one exception, Freeth's mill is the earliest of the kind which we are acquainted with. A specimen was sent to the Society of Arts for exhibition in their museum, but it has disappeared long since. The Freeths were an important people in Birmingham at that time, and the name frequently occurs in the records of the town. As we mentioned in the sixth article of this series, Sampson was connected with Pickard, the Birmingham button-maker, who is alleged to have stolen Watt's idea of applying the crank to the steam engine, and fraudulently patented it. (If there are any descendants of the two families who can throw any light upon what is a very interesting point in the history of the steam engine, we should be very glad to hear from them. It appears from William Hutton's Life, under the date 1764, that Sampson Freeth was one of a little fraternity, ten in number, who were in the habit of assembling together to play fives. His relationship (if any) to the Poet Freeth we have been unable to ascertain.

The date of the introduction of the plated metal trade into Birmingham is given by Mr. William Ryland, in his article in "Birmingham and the Midland Hardware District," as 1764, when it is said to have been commenced by Matthew Boulton, at Soho. The earliest patent having reference to the process is in the name of George Whateley, who describes himself as a "plater of silver," the date being 1768. The object is to obtain gold or silver plated wire and the inventor proposes to plate a round bar of copper or brass from 4 to 5 inches long, and 2 or 3 inches diameter, and then draw it according to the well-known method. If wire of a square section be required a die of a corresponding shape is employed, and flat wire for bullion fringe may be obtained by passing it through rolls. In a second patent, taken out a month afterwards, the inventor claims the plating of gold upon a silver-plated bar and drawing it into wire. Whateley's name does not occur in the list of 45 platers given in the "Directory" for 1770.

It would be difficult to say how far the patent granted in 1770 to Mayer Oppenheim, merchant, for "a new method of making a more beautiful opaque or transparent garnet, or red glass," is indicative of the existence of the manufacture of such glass in the town at that period. He appears in the "Directory" under the phonetic form of "Opnaim," his place of business being 94, Snow Hill. The details of the process need not be given, as the specification may be seen, by those who are interested in the question, at the Free Library. We may mention that a patent relating to the same subject was granted in 1755 to a person of the same name, residing in London. We are probably right in assuming that both patentees are one and the same.

The date of the introduction of what are called "secret" hinges—that is, hinges in which the ends of the pin on which the two flaps turn are concealed—is fixed by the patent granted to John Izon and Thos. Whitehurst, brassfounders, in 1775. The process is very well known, and consists in placing the wrought-iron pin in the mould and casting the metal round it, instead of casting each half separately and uniting them afterwards. The hinge trade was of some importance even so early as the middle of the last century, and in the invaluable old "Directory" for 1770, no less than ten names are given under that head. Izon carried on business in Coleshill Street, but Whitehurst's name does not appear.

That interesting relic of old Birmingham, the Eagle Foundry, Broad Street, has been connected for nearly a century with the industrial history of the town. Portions of the original buildings are still in existence, and some of the old account books of the concern have been preserved to this day. The original proprietor was Richard Dearman, who, in 1776, took out a patent for "an entire new method of making hoes for the American and West India plantations." Nothing but a strong sense of duty could compel us to enter upon the details of this invention, which, we regret to say, is of a thoroughly "Brummagem" character. The hoes are made of *cast-iron*, and, as the inventor observes, "to give them a better body, and bring them nearer the state of malleability, neal them in a gradual heat till that be effected, and afterwards grind them to a proper edge, and harden them." Then we come upon this naïve direction :—"In order to strengthen the eyes of such of them as may be apprehended rather too weak (!) put a ferrule of malleable iron round the same that will enable them to bear the hardest wedging that is requisite to fasten the shaft." This is the most impudent patent we ever came across, and it must be remembered that "malleable" cast-iron could not have been meant, for it was not known for many years afterwards. In 1779 Dearman patented a method of forming the grinding parts of metal mills. After having been cast they are annealed, in order to make them sufficiently soft to allow of the teeth being cut with a file ; the grinding parts are then hardened "with hooves, horns, and bones of animals, or charcoal pounded small ; or common soot mixed with kelp, or common or marine salts laid on with grounds or lees of malt-liquor, or other acid or glutinous matter. Heat them in a furnace, muffle, or open fire, and plunge them in water." These mills subsequently became a specialty of the Eagle Foundry, and there is a representation of one in the pictorial advertisement of the establishment in Bisset's *Magnificent Directory*. Amongst the workmen at the Eagle Foundry — so, at least, tradition asserts – was a certain Jonathan Taylor, who, in 1779, took out a patent for "casting oval bellied cast-iron pots, and nealing, turning, tinning, and finishing the same; and also nealing, turning, tinning, and finishing such kind of round cast iron pots and saucepans as are made with a bead round the top." It is rather difficult to make out what was the precise improvement claimed in this patent, the specification being rather loosely drawn. Cast hollow ware was well known at the time, having been largely manufactured long previously by the Darbys at Coalbrookdale. The method of tinning iron by heating the two metals in contact with the assistance of a flux had also been practised previously, but the application of the process to hollow ware was perhaps a novelty. Some interesting particulars of the rise of the cast hollow ware trade are given by Mr. William Kenrick in "Birmingham and the Midland Hardware District," so that they need not be repeated here. He states that the patent was bought and worked by Messrs. Izon and Whitehurst—mentioned above as the patentees of the secret hinge—who were at the time engaged in the manufacture of cast ironmongery at a small foundry in Duke Street. Simple as it may appear "the bead or rim round the top" was a great improvement, but no special allusion is made to that part of the process in the specification, and it was not practically carried out until many years afterwards by Mr. Archibald Kenrick, of West Bromwich.

In the year 1778 John Talbot, filesmith, patented "a new-invented machine or engine to be used in the working of steel, iron, brass, or copper, hot or cold, in a more expeditious manner than heretofore." At first sight this looks very much like a rudimentary planing machine, but it lacks the distinguishing feature of that monument of mechanical ingenuity—viz., the flat bed which renders the reproduction of a similar flat surface possible. The nature of Talbot's machine will be readily understood when we say that it consists of a pair of rolls, the lower one of which is cut with file teeth. The upper roll runs independently, and is pressed down by a screw or weighted lever. If a piece of metal be now introduced the pressure of the upper roll will tend to draw it forwards whilst it will be ground or filed by the teeth on the lower roll, which the inventor describes as being 30 inches in diameter, and made of wood covered with hard metal capable of being cut with teeth. He states that his machine was "principally constructed for taking the outward coat or crust off any sort of metal;" meaning, no doubt, metal ingots to prepare them for being rolled.

The "scoop saddle," patented in 1780 by Thomas Lindopp, brassfounder and factor, contained a receptacle or cavity "in the nature of a box," for holding—we like to be precise—"two shirts, two pairs of stockings, a pair of shoes, a waistcoat, and a pair of breeches, and also an oil-skin case in the middle or on the side, easily to be come at in case of rain." This is brought about by elevating the seat by a false tree, but in a way which would hardly conduce to the comfort either of man or beast.

William Bache, chemist, who in 1780 patented "a propeller for the use of communicating power to mills, forges, and sundry other important purposes," should serve as a warning to those who dabble in that which they do not understand. It is not very easy to make the specification out, but the "propeller" would seem to be a pump for raising water to turning a water wheel, which may be applied to any purpose where motive power is required. We half believe that the inventor was a perpetual-motion monger, for there is a passage which hints suspiciously at the water wheel as being the means of working the pump.

## CHAPTER XXVIII.

### MISCELLANEOUS INVENTIONS, 1781-95.

The First Gun Patent in 1781—Waters' Bayonet Pistol—William Playfair, of Soho Works—Die-sunk Rolls—Drawing Lath Bars—El Dorado Metal—Patent Buckles—Spring Shields for Ships—Antley's Protected Spur—Strickland's Fire and Burglar Machine—Whitehouse's Patent for Water Plates and Warming Beds—Miles's Cylindrical Wick—Whitworth's Tin-Plating—Whitmore's Mashing Machine—Henry Clay's Coach Panels—Hands' Patent for Plating—Timmins' Patent for Brass Cock Tops—Hancock's Sash Bars—Grove's Coiled Spring Gun Lock-

Continuing our notices of the miscellaneous inventions which have originated in the town, we have to mention the patent granted in 1781 to John Waters, gun and pistol maker, who carried on business at No. 93, Steelhouse Lane, for attaching a bayonet to a pistol. No special method is described, the claim being a general one for the object above stated. This is the first Birmingham patent having reference to fire-arms, the manufacture of which, until comparatively late years, was but little affected by patents.

A passing reference was made in a former article to William Playfair, an old Soho man, whose name appears as a witness to Watt's signature in his copying machine patent. He was a younger brother of the celebrated John Playfair, the distinguished professor of natural philosophy at the University of Edinburgh, upon whom, in consequence of the death of their father, the education of William chiefly devolved. He showed an early taste for mechanics, and was, in consequence, apprenticed to Andrew Meikle, the inventor of the thrashing machine; the well-known engineer, John Rennie, being his shopmate. In common with many of his countrymen he seems to have possessed an instinctive partiality for the south, and he accordingly found his way to Soho, where he was employed as draughtsman for some years. In 1781 he patented "a method of making tongs, spoons, knives, forks, and medals out of solid silver or other metal, which will tend greatly to reduce the price of those articles." The idea of sinking impressions on rolls instead of on dies has been patented repeatedly, as we have already shown, and Playfair's method does not differ materially from any of the others, except that he proposes, under certain circumstances, to use a sector in place of a complete roll, which is to oscillate backwards and forwards, suitable precautions being observed to ensure the correspondence of the two impressions. In 1783 he had removed to London, taking up his residence in Howland Street, Fitzroy Square. He took out two patents in that year for rolling and drawing sash bars, metal mouldings, and beading of all kinds, ornamental metal borders for trays, bottle stands, &c. One of his patents included a method of rolling blanks for horse shoes and for bending the blanks into shape. He used for his sash bars an alloy which he called "El Dorado metal," consisting of copper, zinc, and iron, which we believe to have been identical with "Keir's metal." A large trade was done in articles of the kind by one I. Cruckshanks, of Gerrard Street, Soho, two of whose pattern books are preserved in the library of Sir John Soanes's Museum in Lincoln's Inn Fields. These sashes and mouldings were, in all probability, made by Keir at Tipton, where, as we showed in the fourth article of this series, he had established a manufactory of rolled window sashes of his patent alloy of copper, zinc, and iron, which was capable of being worked at a red heat. Playfair's patent for making buckles and for a mode of plating copper with silver by interposing a layer of fusible silver-alloy between the surface to be coated and the silver are referred to in the article on the buckle trade, which appeared in the *Weekly Post* of May 1. In 1800 he took out a patent for preventing injurious concussions to ships by applying shields fitted with strong springs, so that when a wave struck the ship, the force of the blow was partially dissipated. Possessed of too restless a temperament to persevere with a project for any length of time, he turned his attention to literature, and a long list of his pamphlets is given in Watt's *Bibliotheca Britannica.* He made an attempt to introduce his own inventions, as well as Cort's rolling mill, into France, but political troubles put an end to the scheme. Had he been content to work steadily on at Soho, his name would probably have been as familiar as that of Murdock, for he was undoubtedly possessed of great mechanical abilities. Playfair died in poverty in London in 1823. Particulars of his eventful career are given in a note to Stuart's "Anecdotes of Steam Engines," and in the *Monthly Magazine* for April, 1823.

We have departed somewhat from the chronological arrangement of these articles in the above notice, so as to include all the incidents of Playfair's life in a consecutive narrative. We have now to notice a patent granted to Joseph Antley, spurmaker, in 1786, for a prick spur, that is a straight spike without a rowel, such as one sees in old suits of armour, with a retractile sheath, which in its normal condition always conceals the point. When it is pressed against the horse's side, the sheath is forced back and the spur comes into operation. A spiral spring presses the sheath forward again when the rider's heel is withdrawn from the horse's side. Spurs with protected points were in great favour at the time. In the same year Edward Strickland, who is described as a mechanic, patented "a new invented machine to prevent fire and housebreaking, and which may be applied to different purposes, which will be found of never-failing utility for the protection of the lives and property of his Majesty's subjects." It consists of a complicated arrangement of weights connected to ropes, and so placed that the opening of a door or window tightens the ropes and rings a bell. Fires are indicated in a similar manner, a number of strings being stretched in every direction, and kept in tension by weights. When a fire occurs the strings will be burnt, which releases the weight and rings a bell by the action of a lever. Curiously enough there was published in the town, just at the same time, a pamphlet by William Hamlet, entitled, "The Plan and Description of a Machine calulated to act as a Preservative against Fire and Housebreaking," in which the author, after drawing a gloomy picture of the state of the country, proceeds to describe a machine on precisely the same principle as that patented by Strickland. Hamlet appears to be entitled to the priority, his "Dedication to the Public" being dated November 7, 1785, whilst Strickland's patent was granted in the following June. They may both have been independent and *bonâ fide* inventors, and perhaps the public mind was disturbed by some great robbery or fire which occurred at the time. Panics, fashions, and popular crazes generally leave their mark upon the patent records, and the future historian will, for instance, have no difficulty in tracing the rise, decline, and fall of roller-skating by the aid of the lists of patents.

There is some ambiguity in Isaac Whitehouse's patent, granted in 1787, "for making plates and dishes, substitutes for water-plates, and plates for warming beds, upon a new and improved principle." The articles in question, says the inventor, "are composed of the metal called iron, either of the species called pig-iron, wrought-iron, cast-iron, sow-metal, or iron of such other quality or description as best answers the purpose, reduced into a liquid state, then run or cast in stocks or moulds of different forms, and afterwards covered with tin." A good many patents have been obtained from time to time having reference to that "harmless, useful, and comforting article of domestic furniture," as Mr. Serjeant Buzfuz described the warming-pan on a certain memorable occasion.

The patent taken out by John Miles, gentleman, in 1787, for an improved lamp, looks very much like an imitation of the well-known Argand burner, which was patented three years previously. Instead of a cylindrical wick, he proposes to arrange a number of small round wicks, consisting of a few strands each, in a circle. Another claim in the specification mentions wicks formed of bundles of fine iron wire, the oil rising through the capillary spaces formed by their juxtaposition. This principle has been employed since in what is known as the safety-jet for burning the mixed gases used in the oxy-hydrogen light. In a third modification the oil is raised to the burner by the action of the heat of the flame on a cylinder containing air, which, expanding, presses upon the oil and produces the desired effect, but no means are shown for preventing an undue quantity from being supplied.

The patent obtained in 1790 by John Whitworth, of Deritend, die-sinker, consists of nothing more than plating silver with pure tin, which is to be used for "all sorts of silver-plated goods or wares usually manufactured from gold, silver, or other metals, or metal plated with gold or silver." The inventor illustrates his invention as applied to a knife handle. A plate of silver cut to the proper size is treated in the foregoing manner, and the edges united by soldering. It is then placed in a hole in a block, and a steel plug driven in, which has the effect of forcing the metal into contact with the sides of the hole, and bringing it to shape.

In his poem entitled "The Ramble of the Gods through Birmingham," Bisset takes his celestial visitants

To Whitmore's, then, intent on earth to scan
The wondrous works of still more wondrous man,
They next resorted ; and hydraulics new,
Machines and rolling mills with pleasure view.
* * * * * *
With patent wheels which malt so quickly mashes.

William Whitmore was an engineer and toolmaker of some note, and the mashing machine above referred to was patented in 1792. It consisted of a simple arrangement of gearing, by which a horizontal shaft, fitted with arms for stirring the mash, was caused to revolve, not only on its own axis, but also to travel slowly round the tub, so that no part of the mixture escaped its action. We believe that this was one of the earliest attempts to introduce machinery as a substitute for the hand labour hitherto employed.

Henry Clay, whose name ought to be familiar to every one as the inventor of papier mâché and the founder of the japanning trade, took out a patent in 1792 for a method of "manufacturing panels for coaches and other carriages, also for tables, cabinets, pictures, and other furniture." This was a sort of stone veneer, ground, polished, and painted, gilt, or otherwise ornamented. For convenience of application under certain circumstances it was attached to a backing of wood, canvas, or other suitable material.

In 1794 John Hand, plater, obtained a patent for a method of plating cutlery, by which the operation was performed at a comparatively low heat, thus obviating the risk of injuring the temper of the blade. The knife, having been prepared by filing or grinding, is pickled, scoured, tinned, and anointed with sal-ammoniac. The plate of silver is then folded over it and placed between a pair of tongs, the jaws of which contain heated copper blocks suitably recessed as to hold the blade. When cold it is taken out, hammered, and polished.

The manufacture of cocks must have been in a somewhat primitive condition in 1795 when James Timmins, button maker, took out a patent for securing the plug by a screw and nut instead of riveting it over. The head of the cock is cast in one piece instead of being attached by soldering. This is the first of a long series of patents granted in the town for various improvements in cocks and taps.

The metal sash trade, which has been already referred to in a former part of this article, appears to have attained some importance towards the close of the last century. In 1795, John Gregory Hancock, a famous die-sinker of the town, patented an exceedingly ingenious method of making sheet-metal sash bars, picture frames, fencing, &c., according to which the projecting rib and the ornamental bead or moulding were drawn separately and put together with a bar of metal inside for giving strength, the two parts being connected together by bending over the edges. The construction can hardly be made perfectly intelligible without a drawing. These are probably "Jorden's patent sashes" referred to in the "Ramble of the Gods" above alluded to, one of the witnesses to Hancock's signature in his patent being John Stubbs Jorden. There is no patent in the name of Jorden in the records.

The novelty in Jonathan Grove's patent gun lock (1795) is the use of a coiled mainspring, like that of a clock, instead of the ordinary long bent spring.

## CHAPTER XXIX.

### MISCELLANEOUS INVENTIONS, 1796—1798.

WHITMORE'S ANTI-FRICTION BOX—RUDDER'S CORK-PACKED TAPS—CLAY'S DOUBLE-BODY WAGON—HANCOCK'S EMBOSSED PAPER—THOMASON'S AUTOMATIC CARRIAGE STEPS—HANCOCK'S TWIST-CHAIN—BOULTON'S HYDRAULIC RAM—DECREASE IN PATENTS—HENRY CLAY'S CANAL-LOCKS—CHELL AND NICKHOLLS' BOAT LIFT—FARQUHARSON'S BRICK-MAKING MACHINE.

TOWARDS the end of the last article mention was made of William Whitmore as being one of the ingenious mechanicians who have contributed to build up the reputation of the town. In 1796 he patented a method of preventing the rusting, and lessening the friction of the bearings of weighing machines by enclosing those parts in a box containing oil, a plan which, since that time, has been applied to the bearings of many other kinds of machinery. He also describes a mode of excluding dust from the working parts, an evil to which they are of course particularly exposed. It appears from the *Directory* of the day that Whitmore carried on the business of manufacturing weighing machines of all sorts—a trade which is now of considerable importance in the town.—Following close upon Timmins's patent for cocks, mentioned last week, we have another in 1796, granted to William Rudder, "cock-founder"—a well-established business considerably more than a century ago—for making a groove in the plug for the purpose of receiving a strip of cork or other similar material to serve as packing, and thus prevent the egress of liquid. The drawings attached to specifications of patents relating to these subjects are of much interest, as they show the kind of patterns which were in vogue at the time. They are not very different from those now in use. William Rudder appears in Bisset's *Magnificent Directory* for 1800, as carrying on business with one Handasyd, in Hill Street.

Amongst the ingenious inventions which emanated from the fertile brain of Henry Clay was a double-bodied wagon, for which he obtained a patent in 1796. The object of the improvement was to give the same facilities for tipping the load as those which belonged to the ordinary two-wheeled cart, and the improved vehicle consists essentially of two independent cart bodies mounted on a single frame, so that when loaded the appearance would be that of the ordinary old-fashioned wagon. The body nearest the tail of the vehicle deposits its load at the end, whilst the contents of the other are tipped between the wheels. The contrivance possesses a certain amount of ingenuity, and judging from the fact that it is mentioned in Bisset's book, published some time afterwards, wagons on the improved principle were probably actually manufactured. The remains of one might possibly still be in existence somewhere in the neighbourhood, for although it may seem scarcely possible that implements of the kind should last so long, examples of farm wagons considerably more than a century old were exhibited at the Royal Agricultural Show at Kilburn last year.

The business of embossing paper with ornamental borders or devices is now of considerable importance, and we are disposed to think that the origin dates from the patent granted in 1796 to J. G. Hancock, the well-known diesinker of this town. There is nothing new in the mode of forming the die, or in the actual operation of stamping, the patent being for the application of the process to the ornamentation of paper. Sometimes the die is sunk on a plate which, along with the "force," is passed through rolls—the paper being between—to give the necessary pressure. Stamped paper for revenue purposes was, of course, known and used many years before, and any "American citizen" will give the particulars of a memorable instance in point.

In that very amusing book, "Memoirs During Half a Century," by Sir Edward Thomason, whose house in Colmore Row and factory in Church Street have both been "improved" off the face of the earth, we find an account of his folding carriage steps which he protected by patents obtained in 1796 and 1798 respectively. He says (vol. I, p. 10):—

"I next turned my attention to carriage steps which should move up and down, and fold and unfold themselves, by the action of opening and shutting of the carriage door. Presuming that every physician and surgeon keeping a close carriage would adopt them, I succeeded in bringing my invention to act in 1798. I had the honour, through Mr. Hatchett, to show them to George III., at Windsor; and Messrs. Hatchett and Co., Leader and Co., and Mr. J. Vidler, the three leading coachmakers in London, wrote me letters, expressive of their opinion of the invention, which they permitted me to publish. [We omit the letters as not being of any special interest.] This invention was made public through the medium of circulars. I put a pair to my own carriage, which I used for twelve years. I received orders for many for the Continent, and from physicians in this country; but the great expense of the steps, and the difficulty that coachmakers experienced in putting them to carriages, set the coachmakers against them; and it so occurred, at this period of 1799, that the carriages were principally straight, flat-sided, which form was absolutely essential to the fixing of these steps, but in the following year the doors of the carriages were made of a bulged form, not at all adapted for the peculiar machinery requisite to admit the roller affixed at the bottom of the door to work easily upon the curved lever. I do not think that I made and sold one hundred pair, thereby not realising a profit, but some little loss, by the invention."

We may here allude to an improved chain which, though not patented, was considered of sufficient importance by the Society of Arts to justify the award of a premium of fifty guineas to the inventor, William Hancock, a Birmingham workman. The chain was composed of wire twisted together and then bent, cold, into links, the exact form of which could only be conveyed by a drawing. If the inventor had stopped short before the bending operation he might have been handed down to posterity as the originator of wire *rope*, but this does not seem to have occurred to him. The account in the "Transactions" of the Society of Arts is accompanied by testimonials of a flattering character from J. Miles, William Whitmore, and William Bingley (all of Birmingham), and William Johnson, of Bradley. The chain does not appear to have come into use for coal pits on account of a prejudice conceived against it by the men. Some further information on the subject was given in our "Local Notes and Queries," No. 528, from which it appears that Hancock's chains were manufactured by the William Bingley just mentioned, whose place of business was at Islington.

In the following year, 1797, Birmingham is practically a blank in the patent lists, the only invention protected being Boulton's hydraulic ram. As we showed in the third article of this series, Boulton simply acted as the agent of Montgolfier, who was the real inventor of the machine. What was the cause of this barrenness in a town usually so fertile in invention? Could it have been that the political troubles of the time, as witnessed by the mutiny of the fleet, the restriction of cash payments, and the passing of the Act for suppressing seditious assemblies, were more than usually felt in Birmingham? They certainly had a marked influence upon the whole number of patents granted, for whilst 75 were passed in 1796, of which five belonged to Birmingham, the number in 1797 fell to 54. In 1798 it rose to 77, six of which came from this town. We omitted last week to call attention to the falling off in the number of patents in the year 1791, consequent, no doubt, upon the Riots from the effects of which the town was then suffering. Only one patent was granted in that year, whereas the number in 1790 reached six.

Very soon after the introduction of the canal system the minds of the ingenious began to be exercised to devise some method of preventing the great waste of water at locks. Even the most superficial observer must have been struck by the enormous volume of water necessary to transfer a boat from one level to other. Henry Clay, by a patent granted in 1798, attempted to overcome the difficulty by building locks in pairs, side by side, by which half the water was saved. To make the mode of action quite intelligible, let us take the case of a boat descending the canal. We will distinguish the two lock-chambers as A and B, both of which are supposed to be empty, and the gates leading to the lower level closed. The first operation will be to open the valves so as to fill A, and when this has been accomplished the boat is floated in and the upper gates closed. Under ordinary circumstances the lock would be emptied into the lower level, but instead of doing this, communication is opened between A and B until the latter is half full. The intermediate valve is then closed, and the boat lowered by discharging the half-lockful in A down the canal, and the boat proceeds on its course. The next boat which comes down is let into B, which is already half full, so that it requires just that amount less to bring it to the higher level. Half the contents of B are then let into A, and stored up in a similar way, until another boat is required to be transferred. The plan is extremely ingenious, and if two lock-chambers could be built and maintained at the same cost as one, it would have been a great success.

In the same year, Philip Chell, engineer, and Henry Nickholls, builder, took out a patent for effecting a similar object, viz., the economy of water—but in a totally different manner. According to one part of their scheme, they employed a lift worked by screws assisted by balance-weights for raising the boat and its contents bodily from one level to the other. The second part of the invention consisted in dragging the boat up an inclined plane. The invention is not particularly remarkable, inclined planes having been used on the canals in China from time immemorial, and one was erected at Ketley, near Wellington, Salop, in 1789, the opening of which was commemorated by the striking of a token a short time afterwards. In 1810 a perpendicular lift was put up on the Worcester and Birmingham Canal, at Tardebigg, near Bromsgrove, and a pamphlet, by Edward Smith, illustrated with drawings by W. Hawkes Smith, was published in the town in 1811. A good deal of interest was created by the machine, as is evident from the fact that a third edition of the pamphlet was called for. The lift seems to have worked well for a time, but it was injured by lightning, and was never repaired. All traces of it have been removed years ago.

Amongst the employés in the commercial department at Soho was one Francis Farquharson, and we believe that we do not err in identifying him with a "factor" of the same name who, in 1798, patented a brickmaking machine, which is not only the first of the kind on the patent records, but is interesting as being the earliest type of that class of machine in which each brick is formed separately in moulds on a rotating table. Without describing the machine in detail, it need only be said that the revolution of the table brings each mould successively under the action of a plunger, which compresses the clay in its descent, and forms it into a perfect brick.

We bring our notices of the patents of 1798 to a close with a brief allusion to Francis Hollick's patent for combining an ordinary comb for a horse's mane and tail with a currycomb.

## CHAPTER XXX.

### MISCELLANEOUS INVENTIONS, 1799-1805.

HICKLING'S ENAMELLED HOLLOWARE — CLARK'S PATENT — THOMASON'S GUN LOCK — REDDELL'S AMPHIBIOUS CARRIAGE — GILL'S RIFLING MACHINE — PRITCHARD AND WILLMORE'S PATENT HATS — CHIRM'S PUMP BORER — LONGMORE'S SPRING CLOG — THOMASON'S NE PLUS ULTRA CORKSCREW — TELESCOPIC BRUSH HANDLE AND "FLAGRANTE BELLO" PIKE — DOBBS'S ALBION METAL AND THE CAPSULE PATENTS.

AMONGST the many important inventions which are associated with the town is that of enamelled hollow ware, which was patented by Samuel Sandy Hickling, "gentleman," in the year 1799, the title being "A method of improving and beautifying vessels of cast-iron or other metals used for culinary, chemical, and other purposes, by lining them with vitreous compounds." It is stated in Mr. William Kenrick's article in "Birmingham and the Midland Hardware District," that Hickling established a manufactory for the purpose of carrying out his patent, in conjunction with Mr. Francis, then of the Eagle Foundry, Broad Street. This went on for some time, but the manufacture was eventually discontinued in consequence of a disagreement between the two partners. We have heard another story with regard to this patent, the facts of which are not altogether consistent with Mr. Kenrick's account. It is to the effect that the money for taking out the patent was furnished to Hickling by Mr. Thomas Elwell, of Wolverhampton, who died in 1854, and that the inventor was bound to supply him exclusively with goods as he required them until the value of £80,000. had been manufactured. This was done, a large quantity being exported to Calcutta. These facts were communicated to us by a nephew of Mr. Elwell. A much better process for enamelling was that patented by Thomas and Charles Clark, of Wolverhampton, in 1839, which differed from Hickling's in several important particulars, especially in its freedom from lead. In an action for infringement, commenced by Messrs. Clark against Messrs. Kenrick, of West Bromwich, Hickling's patent was brought forward by the latter as evidence of want of novelty in the plaintiff's process. There is another feature in Hickling's patent which, though curious, can hardly have been of any practical value. It consisted in the use of an alloy of iron and nickel for casting hollow-ware which was coated with copper by immersion in a solution of that metal. If a brass lining was required, the vessels treated as above were to be rubbed over with an amalgam of zinc, and exposed to heat, thus driving off the mercury and causing the zinc to unite with the copper and form brass. We have here what may be termed a "local" application of the old cementation process of making brass. The inventor mentions several articles which may be treated by his method, including, amongst others, fire-irons.

In the same year we find the indefatigable Sir Edward Thomason again to the fore with an improved gunlock, one object of which was to prevent accidents caused by the careless handling of loaded guns. This was brought about by making the cock detachable by the thumb and finger, but as he somewhat plaintively says in his *Memoirs*, "it did not answer, although approved of by several eminent gunmakers—for sportsmen were not aware of any danger in leaving their guns loaded; and I never sold enough to pay my expenses." Another part of the patent consisted of a self-acting contrivance for turning the flint round to a small extent at every shot, so as to present a fresh surface to the striking part and prevent miss-fires.

The patent of Isaac Hadley Reddell (1800) for an amphibious carriage belongs to a very numerous class of inventions which could never have been useful to the public nor profitable to the inventor. The idea was to make a wheel-carriage so constructed as to be capable of acting as a boat when desired, thus avoiding the necessity for trans-shipment of goods which were to be conveyed by road and then by river. Several may be closely linked together, the last one being furnished with a rudder, by which the entire train is steered, like the flotillas to be seen on Indian rivers at this day. As carriages these vehicles must have been very awkward, and they could not have formed particularly handy craft when afloat. Reddell had another patent for coaches and carriages in the same year, dated a few days after the one just described. It embraced a variety of improvements, including modes of constructing and hanging the bodies, the use of an awning to protect the "outsides," making the carriages lofty enough for the "insides" to stand upright, and the seats movable so that the angle of the back might be altered. The latter must have been a great boon in the old coaching days. Besides these improvements there was a ventilator worked by the motion of the wheels, which sucked the foul air from the carriage, the fresh air being passed through water or filtered through fine gauze in dusty weather. Then there was a warming apparatus, an indicator in connection with the wheels for showing on a dial the distance which the coach or carriage had travelled, besides self-acting apparatus for releasing the horses on the occurrence of an accident. A third patent was granted to the same inventor during the year for a stirrup with a spur attached, and so constructed that the foot could not pass too far through and become entangled.

Those who have watched the ingenious and interesting method of cutting the grooves in rifles or large ordnance will no doubt be surprised to learn that the machine dates from the year 1800, the patentee being Thomas Gill, a well-known gunmaker and sword cutler of the town. The barrel is firmly fixed on a reciprocating frame, the cutter —of the width of groove required—being attached to a bar of considerable length, which passes through the barrel and turns in centres at each end. The result of setting the machine to work would obviously be the sweeping out of a *straight* groove from the inside of the barrel as it was forced against the stationary cutter, but if the cutter be rotated on its own axis during the stroke a spiral groove would be the result. The pitch of the spiral, in other words the amount of twist, would depend upon the angle through which the cutter had been made to rotate. If our description is not clear it will no doubt be made so by considering what happens in cleaning a glass lamp chimney. When the brush or other implement is simply moved up and down the dirt is removed in straight lines only, but if it be twisted, corresponding marks will be noticed on the glass. The rotation of the cutters is effected by a rack moving transversely to the bed which takes into a pinion on the end of the cutter bar. The extent of the traverse of this rack, and therefore of the revolution of the cutter and amount of twist given, depends upon the steepness of an inclined plane which forms part of the mechanism. It is not necessary to enter into the details, which are rather for the mechanician than for the general reader, but the specification is well worth a careful perusal. Whether a machine on the principle of the patent was actually constructed at the time is somewhat doubtful, the necessary manipulative skill being then very uncommon. However that may be, the fact remains that a workable machine was designed at a period when mechanical combinations of the kind were but little understood, and so far as we are aware it is the *first* of a large family of similar contrivances. If any of our readers should happen to be in possession of information upon the subject of Gill's rifling machine we shall be glad to receive particulars from them.

It is somewhat difficult to determine the amount of credit which is due to Birmingham in respect to the patent of William Pritchard and Thomas Willmore, dated 1801, for making hats, helmets, &c., seeing that the former was a native of Shrewsbury, the latter being described as a japanner. The process consists in making the articles named by pasting together sheets of paper, canvas, leather, &c. (the first-named substance being preferred) on a block of the required shape. When dry the hat or helmet is dipped in linseed oil, stoved, smoothed in a lathe, and japanned. The process may be used for making the insides of ordinary hats or for strengthening the crowns and brims of hats or caps of any kind.

The patent of Joseph Chirm, pumpmaker, which was granted in 1801, relates to a method of boring pump-trees, wooden water-pipes, &c., and is of purely antiquarian interest. The tree to be bored, after being centred, is suspended vertically between guides, and allowed to descend upon the cutter by suitable mechanism. The drawings show the action of the machine with considerable detail.

The improvement claimed in Josiah Longmore's patent "patten or clog," taken out in 1801, consists in placing a spring in the block or wood of the patten to keep the foot pressed continually against the ties and prevent it from slipping out.

The production of a corkscrew on a sudden emergency has been adjudged by high authorities not to be inconsistent with the profession of strict teetotal principles. We feel, therefore, less hesitation in avowing a particular interest in the improvements which have been made in that instrument. One of the best ever devised was perhaps the "Ne plus ultra," patented by Thomason in 1802. It was certainly the cleverest thing which he ever did in the mechanical way. The peculiarity of the instrument is that one continuous motion of the handle causes the worm to enter the cork and also withdraws it from the bottle. Another advantage is that the cork is discharged from the worm by simply turning the handle in the reverse direction, and the fingers never need be soiled by touching it.

This was an important point, for, as the inventor says in his *Memoirs*, there was then "a kind of fashion for persons to draw the corks of the wine at their own table." It consists of two concentric screws, one left-handed and the other right-handed, the latter being hollow, so as to allow the former to work within it. When the worm has been screwed into the cork to the proper depth, the second screw comes into operation, the handle still being turned in the same direction, and the cork is literally "screwed" out of the bottle by a steady motion, which gratifies the connoisseur by not disturbing the "crust." From a mechanical point of view it is very ingenious, and James Watt paid the inventor a high compliment for the novel combination. The pull is perfectly direct, in a line with the axis of the bottle, as it ought to be, which is not the case with some of the "lever" cork-screws introduced during the last twenty years. Thomason made no less than 130,000 during the fourteen years that the patent lasted. At first they were sold in the London shops at a guinea a-piece, but in the course of ten years the price was reduced to four shillings. They are, we believe, still made in the town, and may be met with now and then in old-fashioned families, who pride themselves on never adopting "new-fangled" things. The Earl of Mountnorris (Lord Valentia) tells how he once heard a native servant at some small town on the borders of the Red Sea call out for "the Thomason" when he wanted the cork-screw. "Thus," says the inventor proudly, "do the Birmingham manufactures find their way into the remotest corners of the globe." The principle of extracting a cork by one continuous motion was also applied many years afterwards by Mr. John Loach, of this town, in a patent granted in 1841.

In 1803 Thomason patented the telescopic hearth-brush and toasting-fork, an invention which, as he tells us, grew out of the method of drawing tubes described in the seventeenth article of this series. His *flagrante bello* pike, which was patented in 1804, consisted in making the weapon hollow at one end, for containing the head, which could be pushed out by a simple mechanism at any moment. This is what would now be called a "sword-stick."

Many of our readers will doubtless recollect the famous case of Betts *v.* Meinzies and others, in which the question of the validity of the plaintiff's patent for making tin-coated lead capsules for bottles was called in question. The patent upon which the defendants relied was that granted in 1804 to Thomas Dobbs, "chemist," of King's Norton, for the manufacture of "Albion metal." The process simply consisted in passing an ingot of lead covered with a plate of tin through rolls with a "hard pinch," so that the two metals were forced to cohere. Another plan was to cast a thin layer of tin on an ingot of lead. When the case came before Lord Hatherley (then Vice-Chancellor Sir William Page Wood), he refused to regard the mere specification as evidence of want of novelty, unless the defendants could prove "prior user," that is, unless they could show that metal had ever been actually manufactured. Diligent enquiry was made amongst the inhabitants of King's Norton with a view to discover if any of Dobbs's old workmen—some of whom were still living at the time—could recollect Albion metal being made. Various applications of this material were mentioned in the specification, including cisterns, vats, distillers' worms, and coffin furniture. The latter offered a very promising field for research, and we believe it was pretty well worked, but no sufficient evidence of prior user was forthcoming, and after a very protracted litigation Betts was eventually successful. In Dobbs's later years he had a rolling mill in Smallbrook Street. He was buried in King's Norton Church. He must not be confounded with James Dobbs, the actor, and patentee in 1814 of a reaping machine. We mention this because Mr. Aitken in one of his articles was under the impression that these two individuals were identical.

The only patent granted in Birmingham in 1805 was that taken out by Richard Kentish, formerly a captain in the Cambridgeshire Militia, for "an armour waistcoat, which is a sure defence against the bayonet, sword, pike, or any pointed instrument, and in many instances may prevent a wound from a musket ball." The armour consists of plates of thin steel loosely attached together by rivets, the holes being enlarged to allow a certain amount of play, and thus conform to the shape of the body, besides offering no impediment to the use of the limbs. It was merely a revival of a particular form of mediæval armour, in the interest no doubt of the numerous volunteer corps in existence at the beginning of the present century. The author published a pamphlet describing the advantages of his invention. Whether he was related to the Rev. John Kentish, who was long well-known in the town, we are unable to say.

# CHAPTER XXXI.

## MISCELLANEOUS INVENTIONS, 1806-1810.

BOURNE'S ROASTING JACK AND STEAMER — PRATT'S TOAST FORKS AND STAND — GOULD'S STEELYARD APPARATUS — WILMOT'S WARMING PANS — NOBLE'S WHITE LEAD — GUEST'S THREAD SPINNER — TOMPSON'S LOCK — NORTON'S FIREPROOF BUILDINGS — GRIFFITH'S AGITABLE LAMP — BLIND RACKS, HANGING GRINDSTONES, COMBINATION WHEELS, CIRCULAR SHEARS, LOOSE BOTTOM CARTS, CHANDELIER DROPS, AND "FIRESIDE COMFORTS."

MUCH ingenuity has been spent on the roasting jack, which has been variously actuated by clockwork, smoke, and lastly steam, but the arrangement patented by Messrs. Bourne, Chambers, and Gould, in 1806, was particularly unpromising. Bourne and Chambers were scale-beam makers of the town, and Gould was an American, who resided for a short time in Birmingham, and then removed to London, where he continued to take out patents. The jack forms a part of a steamer, such as is ordinarily used for cooking vegetables, and the inventors proposed to place within it, immersed in the water, a vane wheel, which should be acted upon and caused to revolve by the steam as it was liberated from the lower surface of the steamer. Motion was given in this way to a pulley which, by means of an endless chain, communicated a rotary movement to another pulley and so turned the joint of meat round. The drawing attached to the specification contains an elaborate representation of the old Staffordshire grate, with the characteristic arabesque ornament on the sides. We refer to this because the pattern has recently come up again along with the Queen Anne revival; but picturesque as it may be, the Staffordshire grate is very inefficient.

In the days when tea was a meal, it was always a great point to keep the toast or muffins hot, and many devices have been patented for holding a plate before the fire. In 1806 Henry Pratt, steel-toy maker, obtained protection for "a new invented toast stand, or an improvement in the article called cats and dogs, upon which things are placed before a fire," It consists of a tripod of twisted rods, made to fold on a central pin when not in use, so that the three rods are parallel, like an artist's sketching stool. Sometimes provision is made to receive a telescopic toasting fork in the middle, so that the whole thing may be stowed away by the side of the chimney piece. There were so many contrivances of this sort that one is sometimes forced to the conclusion that our grandfathers and grandmothers must have been martyrs to portability.

In 1807 Chester Gould, the American just mentioned, patented an apparatus for weighing, based on a principle very much like that of the steelyard, but it would be difficult without a drawing to convey an accurate description. We do not know that it ever came into use, and it does not appear to possess any particular advantages over the ordinary steelyard.

We are not aware that the winter of 1807-8 was particularly severe in Birmingham, but anyhow two patents for warming-pans were recorded. One was the invention of Edward Coke Wilmot, gentleman, who proposed to do away with the danger and inconvenience of using live coals or "gleeds"—we are rather uncertain about the spelling—by employing a heater, which is made hot in the fire and then put into an ordinary warming pan. The other was due to Joseph Johnson and John Wilmot, manufacturers, and was a decided improvement upon the former apparatus, in that it contained a stand for the heater to rest upon. In this way the heater was prevented from coming into actual contact with the pan, the warmth being transmitted by radiation to the top, sides, and bottom of the receptacle which was thus heated more equally. This arrangement is stated to be also applicable for use as a carriage airer. The Wilmot family seem to have had a special faculty for making improvements in warming pans, and we find Edward Coke Wilmot, of Cheltenham, taking out a patent relating to the same subject in 1844. We shall probably not be far wrong in assuming him to be identical with the person bearing the same name just mentioned. If so he presents an interesting instance of long-continued devotion to an apparently trivial question.

The manufacture of white lead has received considerable attention in this neighbourhood from time to time. Kerr's patent has been already referred to, and in 1808 William Moore Noble, surgeon, obtained protection for a method of producing it by acting on lead or litharge with acetate of lead, and then submitting the mixture to the action of carbonic acid. This is the process which was suggested by Thenard in 1801, and is known as the "Clichy process," from having been carried out at that place. It is possible that Noble might have been an independent inventor, but it is certain that the patent contains no features of novelty. A very simlar method was tried on a large scale at West Bromwich many years ago, based on a patent granted in 836 to Gossage and Benson, but it failed.

Old Birmingham directories show that the manufacture of thread was at one time of some importance in the town, and in 1770 there were no less than ten persons engaged in the trade. Amongst them was one Joseph Geast, who carried on business in Digbeth. In 1808, Joseph Mason Guest patented a machine for spinning thread, into the complicated details of which it will not be necessary to enter. In spite of the different spelling of the name, it is probable that both of these persons belonged to the same family, as "Geast" is often pronounced like "Guest," and there may perhaps have been a clerical error in the specification. It has been suggested that the thread manufacture was founded by Paul and Wyatt, whose experiments were fully described in the second article of this series. It is now quite extinct here.

The patent of William Tompson, of Deritend, locksmith, which was granted in 1808, is sufficiently explained by the title, which is "a new-invented lock, which acts in a perpendicular and then in a horizontal direction, with springs and tumblers, one part being at liberty whilst the other is in motion, the bolts of which lock return into the body thereof when it is unlocked."

Besides those already referred to in former articles under their proper head, three patents granted in 1809 remain to be noticed. William Wells, of Erdington, cock-founder, obtained protection for a method of preventing leakage in cocks by casting grooves in the plug to receive packing—an old contrivance even at that early date. Charles Norton, builder, had a patent for constructing fireproof buildings, which consisted in the use of brick arches in combination with iron girders. The drawings show various forms of beams and roof-principals which the inventor proposes to employ. The slates are fastened by nails driven into wooden plugs let into recesses provided at various points. This patent is of interest as showing the views which then prevailed on a question which has since assumed considerable importance, but the girder with a top and bottom flange, now in everyday use, was not known at the time. Edward Griffith, optician, patented in the same year, "an improved air-tight agitable lamp," which consisted of a hollow stem containing oil, having a small bucket fitted to slide in the cylinder. The bucket was pushed down into the stem at intervals and brought up with it a supply of oil for a short time. It was kept closely pressed against a flange at the top so as to prevent the principal reservoir from being upset, but there appears to be no provision for preventing the liquid contained in the bucket from being spilt.

The patents of 1810 included one by Thomas Bayley, brassfounder, for that provoking article of window furniture, the blind-rack, which has exercised the ingenuity of inventors for more than half a century. Bayley's idea was to fix the pulley by a screw, instead of the ordinary rack and spring. John Slater's method of hanging grindstones consisted in using two circular plates which were screwed tightly against the sides of the stone, very similar to the plan now employed for fixing circular saws upon their spindles. The evil effects of driving wedges into the eye of the stone, which had a tendency to burst it, was thus avoided, but the mode of obtaining the pressure was rather different to that now used in the mounting of a circular saw. William Doughty's "combination of wheels for obtaining motive power" looks, at first sight, very much like Watt's "sun-and-planet" motion, and the principle is in truth very similar, but mechanicians will understand us when we say that it is nothing more than the "epicylic train" which was very well known at the time, as it had been frequently employed, especially by the French, in the construction of equation clocks. The patent of William Bell, of Handsworth, engineer, for a machine for cutting pasteboard, &c., consisted in arranging a series of pairs of circular cutters on shafts, the cutters being adjustable so as to cut various widths. This principle is now in very common use in a variety of trades. John Bosworth's cart had doors in the floor of the vehicle, so as to give greater facility in discharging the load. The doors were hinged at the sides, meeting in the middle, and were opened and closed by means of a chain wound round a roller at the side. The next patent on our list contains two well-known Birmingham names in connection with the glass trade. It was an invention for fixing wire loops in glass drops for chandeliers by William Shakespear and Thomas Osler, the younger, "glass chandelier furniture manufacturers." Allusion has already been made to the contrivances in use for promoting conviviality at the tea table, and Benjamin Ager Day, of Handsworth, brassfounder, had a patent for various "fireside comforts," as he calls them, including a toast stand, fitted with a bracket for screwing to the fender, a telescopic hearth brush, and a portable toasting fork, the prongs of which are retractile and are drawn into a sheath when not in use, and slidden into a hollow in the stem of the hearth brush. Our forefathers used to be very much troubled about their snuffers, for the improvement of which many patents were granted. William Hobday proposed to do away with the inconvenience of springs, which were always getting out of order, by substituting an arrangement of toothed wheels and levers, which latter have somewhat the appearance of the mechanism of a gun-lock, except that there is no spring—a point upon which the inventor particularly insists. The amount of ingenuity which has been lavished upon snuffers is remarkable, considering the extreme simplicity of the operation to be effected. But inventors were ambitious, and they endeavoured to produce an instrument which should not only cut off the superfluous wick, but should also retain it securely in a chamber, thus preventing the snuff from being dropped about when the snuffers were opened for a fresh operation.

## CHAPTER XXXII.

### MISCELLANEOUS INVENTIONS.—1811-1812.

STEAM PROPELLER FOR CANALS—CUTTING DIES FOR TRIMMING CAST WORK—BLIND RACKS—"SHIELD GLAZING" FOR HOTHOUSES—SAFETY TAPS—STEEL-IRON TOOLS—STEAM ROTARY ENGINE OR PUMP—BRASS-COVERED IRON TUBES, &c.—DOUBLE STEEL-YARDS—ECONOMIC LAMP AND WICK CARRIER—NOVEL SWORD SCABBARDS—B. COOK'S PATENTS FOR CASED IRON TUBES AND RODS.

THE inland position of the town, and the want of a navigable river—we apologise to the deity of the River Rea—does not seem to have prevented the men of Birmingham from turning their attention to steam navigation. In 1811 Henry James took out a patent for propelling vessels by steam, the driving mechanism being a paddle-wheel fixed at the stern, in the centre line of the ship. The vessel was steered by two projecting vanes or "fins," placed one on each side near to the stern. The invention is stated to be particularly applicable to canals, and especially for taking boats through tunnels, as a substitute for the labour of those unfortunate beings known as "leggers," whose function it is to lie flat on a plank, and propel the boat by pushing with their feet against the roof of the tunnel. Not only is the work particularly severe, but it is attended with the risk of rolling off into the water and being crushed between the boat and the wall. The profession of the patentee is not given, so that it is not possible to identify him with a person of the same name who patented an improvement in gunmaking machinery, in conjunction with John Jones, a few months later. The patent is of interest on account of the date, as it preceded Henry Bell's successful attempt to establish a regular system of steam communication on the Clyde by about a year and a half.

William Jenkins's patent, of 1811, "for removing the surplus metal from cast "flat-backed handles and rings," by means of a stamp instead of filing, gives a curious insight into the state of the cabinet brassfoundry trade seventy years ago. The "fin," or projecting ridge of metal caused by the joint of the mould, was usually removed by the file, which was not only an expensive but a wasteful process, much of the metal being lost. Jenkins proposed to remove it in one piece by the action of a cutting die.

Reference was made last week to the numerous contrivances which have been proposed to obviate the almost perennial nuisance of badly-made blind racks. Ralph Sutton's plan was to keep the cord always at the right tension by connecting the bearings of the small roller round which the cord passes to a spiral spring. By a simple arrangement of parts which it is easy to imagine, the spring was constantly pulling down the roller, thus compensating for the stretching of the cord with use, and keeping it always tight. There have been many modifications of this plan, which is a very good one, and answers well as long as the spring retains its power, but when once overstrained the thing is useless. Sutton was a brassfounder, and his patent is dated 1811. He appears to have been an importation from Macclesfield; at any rate, a patent was granted to a "brazier and tin-plate worker" of the same name who was residing there in 1806 for a method of cooking by steam or hot water. He also took out another patent, which we shall notice presently.

Jorden's window-sash manufactory is celebrated in Bisset's "Ramble of the Gods through Birmingham," but he does not state where it was situated. His patent of 1811 was for a method of glazing horticultural buildings and skylights, the peculiarity of which consists in a provision for the escape of the moisture deposited by condensation on the inside of the glass. He calls it perforated "shield glazing." The panes, which are shield-shaped, overlap, and are cemented together by a composition not described, a small aperture being left at the base of the shield for the escape of the products of condensation. We are inclined to think that the William Walker Jenkins, brassfounder, who took out a patent in 1811 for drawer-knobs, must be identical with the William Jenkins referred to above. His improvement consisted in using tin-plate as a material for covering the iron bodies of knobs. It would thus appear that the rage for cheapness was affecting the trade even at that early date.

The year 1811 closes, so far as Birmingham is concerned, with a patent granted to William Rudder, cockfounder and warming-pan maker, for a method of keeping the plug of a cock tight by a screw at the bottom. It is scarcely distinguishable from the method which was patented by James Timmins in 1795, and which we alluded to in our article of September 4, as indicating that the beer-cock of the day was a very primitive contrivance, the plug being held in its place by riveting the bottom of it over. Another part of Rudder's invention consists of a plan for preventing cocks from being surreptitiously turned. A collar forming one piece with the shell has a pointed screw fitted into it, and in the plug is a corresponding recess to receive the end of the screw above-mentioned. When the screw is driven home, it is obvious that the plug cannot be turned; but these "night cocks" offered no obstacle to those who were possessed of an ordinary screw-driver.

The year 1812 is noteworthy as having given rise to a larger number of patents than were ever granted in the town in any previous year. True, there were only 11, but then the entire number issued during that period amounted to no more than 119, 53 of which belonged to London. The fact that more than three times the number of patents were applied for from Birmingham as compared with any other provincial town would seem to show that the spirit of invention was very active here. Three of the more important patents have already been dealt with; but we have to notice one which was granted to George Babb, of Bordesley, engineer, for making fire-irons, files, and edge-tools, by converting them into steel by the ordinary process after they have been forged to shape. The parts not required to be steeled are bedded in clay and road earth. The inventor describes an improved form of furnace which he prefers to use, and he states that the work does not blister, and that less time is required than with the ordinary steel-converting furnace.

Hundreds of inventors have spent their strength in devising a steam-engine somewhat on the principle of the windmill—that is, one in which the rotation shall be continuous. As we have already seen, it was tried by Murdock, and also by Watt, and in 1812 Philip Chell, engineer, took out a patent for a similar machine. The patentee states that it may also be worked by a current of water, with a few modifications, and it did not escape him that by reversing the action of the machine it might be converted into a pump—a property possessed by all contrivances of the sort.

The name of Loach has long been known in the brass-foundry trade, and in 1812 Mr. John Loach took out his first patent for castors and lock furniture, the principle of which consisted in covering such articles with sheet brass instead of making them solid. This might have been suggested by the great success of Cook and Attwood's patent for brass-cased iron tubes or rods, which was granted in the previous year, and which was described in a former article.

In spite of its formidable designation, William Whitfield's patent "one-side compound lever steelyard" was a very simple contrivance, the object being to give a wide range to the apparatus, and adapt it to weigh heavy as well as light articles, without increasing the length of the beam, or using a different sliding weight. This was accomplished by having two hooks for suspending the steelyard, one being placed much nearer the end of the beam. When it is required to ascertain the weight of a very heavy body the apparatus must be suspended from the hook nearest the centre of the beam. When, on the other hand, the object is small the hook nearest the end of the beam is used. It will be understood that two distinct graduations are necessary, and accordingly those for use when the short arm is employed are cut on the front of the beam, whilst another set for indicating the weight when the long arm is brought into play are marked on the top of the beam. The principle of the invention will, perhaps, be more readily understood by considering the case of the ordinary scales with equal arms, which require a weight equal to that of the object to be weighed. If now, the point of suspension be moved nearer to the pan containing the body whose weight it is desired to ascertain, it will be obvious that a weight of (say) one pound will be sufficient to balance one of four, five, six, or more pounds, according to the relative distances between the point of suspension and the scales. Whitfield's patent was in effect for a steelyard, the length of the short arm of which might be varied. It was probably intended to supersede the compound steelyard made after the old Roman pattern, in which the variation of the length of the short arm was effected by reversing the mode of suspension.

The patent of Bassett Burrows, hatter, for an improvement in waterproof hats, consists in applying a coat of shellac varnish on the inside of the hat. It would be interesting to ascertain why John Simpson, who is described as a tin-plate worker, called his patent lamp "Palmer's Birmingham Economic Lamp." The well-known Palmer's candle lamps did not come into being for nearly thirty years afterwards. Simpson's patent chiefly related to the form of the reservoir, and to the use of deflectors for supplying air to the flame. These deflectors were also made to serve the purpose of snuffers, the carbonised wick being removed by causing them to sweep over the top of the burner. An improved wick carrier is also described. William Deakin, wire-drawer, of Deritend Mills, appears in this year (1811) as patentee of a method of making scabbards and sheaths, the essence of which appears to be the covering of metal scabbards with leather, cloth, mohair, paper, or any similar material which will serve as a ground for holding varnish. The articles may be lacquered or painted in any desired manner.

As we remarked just now, the patent for casing iron tubes or rods, or wood rods for cornice poles, was taken out in 1811, and the manufacture appears to have been rapidly developed. In 1812 the ingenious inventor Benjamin Cook, who well deserves a prominent place in a "gallery of industrial worthies" of the town, obtained a second patent for a great variety of applications of the material. As Cook was an eminently practical man, we may take the specification to indicate in some measure the kind of articles which were in fashion at that time. He proposes to use wood cased with brass for the frames and laths of dwarf Venetian blinds—which, by the way, were the invention of Gowin Knight, a celebrated philosopher, well known for his magnetic researches; and he also mentions door-frames, window frames, carriage-wheels, cornices, picture frames, frames for looking-glasses, bedsteads, furniture, and ballusters for staircases. He applies iron cased with brass to the manufacture of the pillars of fire-screens, bedsteads, and he also proposes to make military spears, to fold up like a telescope, of the same material. Venetian blinds with metal laths have been frequently brought out as "the very latest novelty," but Cook seems to have been the original inventor, and they are included in this patent. He proposes in some cases to use perforated metal for making the laths, both of the ordinary and the dwarf kind, which he says will produce a very pleasing and novel effect.

# CHAPTER XXXIII.

## MISCELLANEOUS INVENTIONS, 1813-1816.

PATENTS FOR HOUSEHOLD FURNITURE — SLIDING FENDERS — WINDOW SASHES — BRASS CHIMNEY PIECES — SOCKET CASTORS — RIDING WHIP SPUR — STEAM WASHING — STEAM HEATING AND COOKING APPARATUS — WATER-PROOF GUN LOCKS — AXLE BOXES—IRON FENCING—ACTORS' PATENTS—DOBBS'S REAPING MACHINE—STRAW CUTTING MACHINE — CANAL LOCK—CUTTING WIRE FROM PLATES—SMOKE CONSUMING FURNACE — GLASS DROPS FOR CHANDELIERS—CURIOUS THIMBLES.

THE first Birmingham patent granted in 1813 was that taken out by Thomas Ryland, plater, for "a fender on a new construction, the principle of which is applicable to other purposes." It consists in making fenders with an additional rail to be elevated at pleasure so as to increase the height of the front, which is accomplished by fitting the rising front with telescopic sliding tubes. The "other purposes" mentioned in the title are nursery guards and window blinds.—Following this we have Charles Plimley's patent for rolling taper bars (both round and square), file blanks, bayonets, matchets, swords, ramrods, &c., which he accomplishes by the oft-repeated device of cutting taper grooves corresponding to the dimensions of the articles to be produced in a pair of rolls, through which the metal is to be passed. When dealing with very long bars, the grooves are cut helically round the rolls. Suitable means are provided for keeping the corresponding parts of the two rolls always in the same relative positions. The inventor also mentions that one roll may be plain, or may even be replaced by a flat plate, the recess in the roll being in these cases cut just double the depth.—In the same year James Timmins, manufacturer of window sashes, invented an improvement in hothouses, horticultural buildings, and forcing frames, which consists in covering wrought iron bars of suitable section with sheet copper or brass, the metal being first stamped into a trough-like section, corresponding to the outline of the bars. The details of the mode of erection need not be described, the object of the invention being to prevent the corrosion of the iron, and by using wrought instead of cast metal to considerably reduce the weight, and also offer less obstruction to the light.—The patent of Robert Lewis, brassfounder, was for the manufacture of brass chimneypieces, the various portions being cast separate and screwed together. They are attached to the chimney fronts by means of bolts and nuts, the principal object of the invention being to give facilities for their easy removal without disturbing the brickwork. The ornamental parts are sometimes made separately either by casting or stamping, and attached by screws or springs. William Jenkins, brassfounder, patented an improvement in the manufacture of socket castors, consisting in "making some with raised sockets, others with stamped sockets, and others with sockets both raised and stamped, all of which sockets by such my improved method being made whole, and not in parts." The sockets were raised either at a press or stamp.—Opinions will probably differ as to the value of Sir Edward Thomason's patent riding whip, which was fitted with a spur-rowel, or spike, at the end, an arrangement similar to the slide of a pencil-case being provided for protruding the sharp points whenever it was desired to remind the horse of his duty.—Ralph Sutton, who has been already mentioned in these articles, patented in 1813 a method of preventing the accidental discharge of fire-arms by fixing a flat ring in a recess in the barrel, so as to cover the touch-hole. There is a hole in this ring of the same size as the touch-hole, and when it is desired to fire the piece the ring is turned until the two apertures coincide, so that the fire strikes through when the trigger is pulled.

The patents of 1814 include one granted to John Slater, for washing clothes and warming buildings by steam. The washing apparatus consists of an ordinary furnace, with a steam-tight lid, steam from which passes into the vessel containing the clothes. "Hot closets," as they are called, are provided for drying clothes, and are heated by "a frame resembling a common clothes-horse, but made of hollow rails or tubes into which steam is conducted." The inventor also describes a "coil" for heating apartments, and he accomplishes the same end by passing steam into chimney-pieces, or furniture made hollow for the purpose. In the *Monthly Magazine* for 1811 there is some correspondence relating to Slater's steam cooking apparatus, but nothing of the kind is described in any of his patents until the year 1821.—Although we have not been able to trace the relationship of Emanuel Heaton, gun finisher, who took out a patent for a water-tight pan for gun locks, in the year 1814, we have no doubt that he belonged to the well-known Birmingham family of that name. The improvement applied, of course, to the old flint lock, and simply consisted in placing a packing of leather round the edge of the pan which contained the priming.—The object of the patent for an improved axle-box, granted to William Whitfield, scale-beam maker, whose improved steelyard was described last week, was to lessen the friction of coach and carriage axles and the bearings of machinery, by a peculiar arrangement of receptacles for the lubricant. The details include certain features which have been long in use, but they are too complicated to be described here. —The patent of William Neville, merchant and coach-founder, seems to mark the date (1814) of the invention of hurdles, gates, palisades, balustrades, &c., with hollow instead of solid bars. The drawings attached to the specification show the various methods by which the bars may be connected together.—J. S. Jorden, who has been already mentioned as a manufacturer of sash frames and bars, took out a patent for what he calls a "waterscape bar," to be applied to rafters, skylights, shop fronts, windows, and, in short, to any part of the interior of a building where moisture is likely to condense. It consists of a kind of trough which serves to conduct away the moisture which is deposited by condensation.

The number of actors who have been patentees is very small, and with the exception of the individual about to be mentioned we only remember one. And, even here, we are not quite certain whether the Dion *de* Boucicault, who took out a patent for irrigating land, in the year 1849, is identical with the well-known actor and dramatist. James Dobbs was a comedian and song-writer, greatly esteemed in the town, and was a nephew of the Thomas Dobbs, referred to a few weeks ago, who had a rolling-mill at Lifford, King's Norton. Several particulars of his life appeared in our "Local Notes and Queries," last autumn, so that they need not be repeated here. He appears in these articles by right of his patent for a reaping machine, which he exhibited in action on the stage of the Theatre, October 14, 1814, an artificial field of wheat having been planted for the occasion. He took the part of *Robin Roughhead*, in "Fortune's Frolic," the reaping machine being introduced as a sort of interlude. The experiment was fairly successful, but we hear nothing of the machine coming into actual use. The cutting mechanism consisted of a circular saw, which the inventor calls a "circular sickle," the grain being fed up to it by means of a pair of rollers, which draw the straw in between them. In some cases the inventor proposes to use circular shears, such as are largely employed in manufactories. Here he was to some extent anticipated by William Bell, whose patent for a straw-cutting machine on the same principle was described a few weeks ago. The machine was capable of cutting not only corn but beans, maize, rice, tobacco, and sugar cane.

The eventful year 1815 was singularly sterile in invention, so far as the town is concerned, only two patents having been granted. The first was in the name of Thomas Bagot, surveyor, for an improved canal lock for economising water, which can hardly have survived the experimental phase, if, indeed, it ever got so far. The lock has three chambers arranged side by side, the central one being fitted with a "vacator," as the inventor calls it, but which is in fact a huge piston, kept in water-tight contact with the sides of the lock by means of a leather packing. Instead of allowing the water from the side chambers to pass away into the lower reach of the canal it is let into the central chamber and forced as required into one of the side chambers by loading the "vacator" with weights, or by the action of a hydraulic press.—The second patent was that granted to William Bell, for making wire by rolling, or rather, slitting a flat plate. The rolls employed have semi-circular grooves with sharp edges which cut the plate into strips and fashion it into a circular form. The process is precisely similar to that patented by W. E. Sheffield in the previous year.

In 1816 five patents were granted to the town; two have been already described, and of the three which remain the first is that taken out by Mr. G. F. Muntz for a smoke-consuming furnace. The apparatus was based on a principle which has since been effectually applied to the condensation of the vapours proceeding from copper and lead smelting furnaces and alkali works. It consisted in passing the smoke through a shower of water, produced in any convenient manner. Some provision had to be made for reinforcing the draught, which was naturally much diminished by the cooling effect of the water.—In the same year John Aston Wilkes, glass manufacturer, took out a patent for attaching glass drops to chandeliers by means of an eye, to be made separately by winding a filament of softened glass round a wire. The eye may also be formed by piercing a hole through a solid disc of glass and rounding the edges in the flame.—The patent of John Piercy, tortoiseshell toy manufacturer, for making thimbles, consists in combining in various ways the use of metal, either as an inside or outside lining, with tortoiseshell, horn, leather, wood, &c. Five different combinations are described, but there is no claim for any special machinery.

## CHAPTER XXXIV.

### MISCELLANEOUS INVENTIONS, 1817-1819.

ART MANUFACTURES—PSEUDO-GOTHIC ORNAMENTS—DOUBLE AND ELASTIC-STEELED REINS—GOLD ON COPPER PLATING—SPRING HOOKS—BLOWING ENGINE—SPRINGLESS CANDLE SNUFFERS—SAFETY HARNESS—ROLLED FORKS AND SPOONS—"POWER MACHINERY"—MECHANICAL OARS—CURIOUS STEEL PROCESS—EARLY KITCHEN RANGE—WILLIAM BRINTON, OF SOHO, AND HIS INVENTIONS.

ABOUT thirty years ago the department of Practical Art, whose headquarters were then at Marlborough House, organised a sort of artistic "Chamber of Horrors," consisting of all the "frightful examples" of bad design in carpets, paper hangings, and other domestic adjuncts which could be collected. It was a very striking exhibition, and although it contained many articles of conspicuous ugliness, there was yet a lower depth which was almost entirely unrepresented. We allude to the "Gothic" patterns which were in vogue about half a century ago, drawings of which may be found in any old brassfounder's pattern book. We are reminded of them by the patent of Benjamin Ager Day (1817), for "certain improvements in chimney ornaments, which said chimney ornaments are so constructed that they may be used for fire-screens, flower or scent jars, time-piece cases, candle sticks, toast stands, and various other purposes." The mode in which these articles are produced is legitimate enough, consisting as it does in making them in halves and uniting by soldering or brazing. The ornaments are "to represent Gothic or Grecian architecture," and the drawings attached to the specification exhibit in a very remarkable way the meaning which was attached to those words at the time. We have, for instance, a hideous clock-case, in shape something like a mediæval shrine, but in which Classic and Gothic forms are hopelessly intermixed. Crocketted pinnacles are surmounted by a well-known Greek ornament, whilst an acanthus leaf—made to serve as the base of a column—is in close proximity with something which is perhaps intended for an imitation of the triforium arcade of Westminster Abbey. On the ridge of the roof of the structure there is an aperture, something like a chimney, for holding a small hand screen, which is of peculiar construction and made so as to fold after the manner of a fan. According to another part of the invention a toast-stand is so made as to be suitable for use as a chimney ornament.

Notwithstanding the importance of the saddlery trade the patents relating to the subject which have been issued to residents in the town are not very numerous. Thomas Taft, who is described as a saddler, whipmaker and bridle-cutter, patented, in 1817, an improvement in bridles, which he describes as follows : "I divide a single rein, at any distance in the rein, so as to give a double purchase, or in other words, to leave two or more reins on each side the horse's neck, to be fastened to a bridle, bit, or other thing, and the rein only for the hands ; and this I do either by splitting, cutting, coupling, or sewing." The invention also mentions reins lined with elastic steel for military use, so as to prevent them from being cut through by a sword, and he proposes to make reins of twisted catgut, covered with whipcord, in the manner that whips are made. In the same year Mr. John Turner, button-maker, specified a process for plating gold upon copper or brass, by placing a sheet of the first-named metal upon an ingot and hammering the plate of gold until the two are in intimate contact. The whole is then to be bound round with wire, and the junction of the two metals covered with a mixture of silver filings, or solder and borax. Heat being applied the plate of gold becomes effectually attached to the ingot.

The first patent of the year 1818 is in the name of John Simpson, plater, for spring hooks for harness, and consists in applying the spring at the back of the joint on which the movable part turns, and lodging it in a hollow or cavity so that it will be protected from dirt. Although steam engines are expressly mentioned in the title of the patent granted to Thomas Jones, of Bradford Street, iron-founder, and Charles Plimley, refiner, the improvement is only described with reference to blowing engines. Instead of the usual packed piston, the inventors propose to employ a hollow cylinder, closed at one end, which works in a fluid packing like an ordinary gasholder. When the receptacle is forced downwards it is obvious that the air within it will be compressed, and if a suitable eduction pipe is provided the air will be forced out. When it has arrived at the end of its stroke, the motion is reversed, and a fresh supply of air is obtained through a valve arranged to open inwards, the eduction valve being closed. It will be seen that this plan would only be applicable where a low pressure of wind was required, as the tendency would be to blow the liquid packing out of the cylinder.--The peculiarity of Samuel Hobday's snuffers is that they have no springs or levers, the various parts being actuated by inclined planes upon the handles.—William

Finch, gentleman, patented a hollow bit for horses, the object of which was to convey water or any other fluid into the mouth by means of tubes attached to the ends, which tubes were conducted upwards and fastened to some part of the harness, A sponge is placed in the lower part of the tubes to regulate the flow of the liquid.

Of the eight patents which originated in the town in the year 1819, the first is that granted to John Simpson, plater, for "release harness," the object of which was to provide a speedy means of letting a runaway horse go free, or in case of accident to the carriage. Hundreds of contrivances of the sort have been patented, but horses still continue to run away and overturn the vehicle to which they are attached, The present specification includes several varieties of connections, but they need not be described. The patentee alludes to his improved buckle patented in the previous year, and described above.—Samuel Haycraft's patent for making spoons and forks consists in passing the metal out of which the blanks are to be cut through rolls which are so constructed as to press the metal into plates of varying thicknesses, answering to the thickness of the various parts of the article to be made. The blanks are cut from these sheets in the ordinary manner, and brought into shape by stamping between dies. When required to be plated they are finished in another pair of dies.

It is in the class of patents for motive power that one finds the most extraordinary examples of perverse ingenuity—of elaborate contrivances to bring about a result which may be far more easily accomplished by simple every-day expedients, William Neale, whitesmith, has such a patent for "an improved combination of machinery calculated to increase power," which consists in converting motion by the agency of what is known as a "lazy-tongs" arrangement, in combination with levers, connecting rods, and cranks. According to another method he employs a mangle rack and other mechanism to accomplish the same object. —James Mason, merchant, obtained protection for a method of mounting oars for rowing ships, which does not seem to promise much advantage. The shaft and blade are in two parts, the former being mounted in bearings on the boat's side, and connected by links to the blade, which is fixed at a lower level. The motions of the handle are exactly reproduced in the blade, just the same as with an ordinary oar, the only difference being that the axis upon which the blade part works is much lower down, and nearer the surface of the water. The inventor says that the power used may be that of horses, steam, or wind, but he gives no description of a mode of applying such forces.—

So far as we are able to understand the patent granted to Stephen Bedford, ironfounder, of Birchall Street, it consists in converting iron into steel by heating it (in a manner similar to the ordinary cementation furnace) in contact with "vitrified iron, slag, scoria, cinder or other denominations of the refuse of iron, or iron ore." The iron may be in bars, or fashioned into any of the articles usually made of steel; but it is not easy to follow the patentee in his description of the effect which takes place, especially when he goes on to say that he improves the quality of other metals by heating them in the same manner in contact with "their respective vitrifications, slags, scorias, cinders, or other denominations of the refuse produced in the smelting or fusion of each respective metallic ore." There is some mistake or intentional ambiguity about this specification.—The numerous improvements in cooking ranges and stoves which have been introduced during the last quarter of a century must not make us forgetful that a Birmingham joiner, one John Heard, contrived a very perfect apparatus of the kind more than sixty years ago. The drawings attached to his specification show a neat, compact, and well-designed range, standing quite free of the wall, and capable of performing the various operations of roasting, boiling, baking, besides being adapted for warming the apartment in which it is placed. The products of combustion are carried off by a flue leading into the chimney of the room. The inventor mentions that it is particularly suitable for use on board ship.

Amongst the patents of the year 1819 there is one in the name of William Brunton, an old Soho man, whose memory should not be allowed to perish. Like many of his colleagues at that famous establishment he was a Scotchman, having been born at Lochwinnoch, Ayrshire, in May 1777. His father was a clock and watch maker, a trade which then engrossed all the minor mechanical works of the district, and in his shop, whilst yet a child, he laid the foundation of his future mechanical proficiency. In 1790 he was put to work in the fitting shops of the New Lanark Mill, then recently erected by Mr. David Dale, in conjunction with Sir R. Arkwright. Attracted by the fame of Soho, he migrated southward, and in 1796 he was taken on there. So rapid was the progress that he made in the good opinion of his employers that, when he was only twenty-one, he was sent alone to the collieries of Mr. Curwen, in Westmoreland, to rectify some defects in a pumping-engine erected there by Boulton and Watt. Of the difficulties which he had to encounter in this his first campaign he used to speak with great interest to the latest period of his life. He remained at Soho until he was made foreman and superintendent of the engine manufactory, and was thus brought into close intercourse with Watt, especially during the later years of his active work as a partner. Brunton used frequently to recur to his recollections of Watt at this period, to the unrepining manner in which he bore his sufferings, and the interest he took in all matters relating to the engine business, even when disabled by illness. Brunton left Soho in 1808 to become a partner in the Butterley Ironworks; but he was back again in Birmingham in 1815, when he took a share in the Eagle Foundry, of which he was the mechanical manager. He remained there for ten years, during which period he designed and executed many important works. In 1825 he removed to London, and commenced practice as a civil engineer, and ten years later he became a partner in the Cwm-Avon Tinworks, where he erected copper and smelting furnaces and rolling-mills. He died October 5, 1851, at Camborne, Cornwall, at the house of his son William, having some years previously lost the whole of his fortune in an unsuccessful commercial enterprise. Brunton took out nine patents in all, three of which were granted during his residence in Birmingham. The first was for a steam-boiler furnace, with a revolving bed, the fuel being fed through a hopper to which a tremulous motion is imparted, so as to gradually distribute the fuel evenly over the bed as it is rotated beneath, The specification contains no drawing of this ingenious contrivance, but an illustration made from an actual furnace will be found in Partington's "History of the Steam-engine." The patent also contains a description of a piston having an internal annular reservoir for containing oil, tallow, or fluid metal. The lubricant has a free communication all round, so that no steam can escape whilst any of the fluid remains in the recess, He took out patents for improvements in his furnace, in 1820 and 1822 respectively; but the invention by which he is best known is the walking machine or "steam horse," which he brought out in 1813, when he was at Butterley. At that period engineers were under the impression that the adhesion between the driving-wheel of a locomotive and the rails would not be sufficient to cause a sufficiently powerful "bite" when drawing heavy loads. Taking the human legs as his model he constructed a machine in which the action of the various parts of the limb was exactly imitated. It was very successful, and during the winter of 1814 it drew coals up an incline of 1 in 36, at the Newbottle Colliery. The career of the "steam horse" was, however, brought to a sudden and painful conclusion, the boiler having exploded and killed thirteen persons. His remaining patents had reference to calcining furnaces, machines for dressing ores, excavating machines, &c., but none of them were the source of much emolument to him.

## CHAPTER XXXV.

### MISCELLANEOUS INVENTIONS 1820-1823.

KNIFE HANDLES—TABLE CASTORS—GOLD AND SILVER RAZORS—TANNING PROCESS—COOKING RANGE—MEAT JACK—DOUBLE SNUFFERS—LAMP BLACK—DETONATING GUNS—FIRE-PROOF WOOD—SMOKE CONSUMING—PIANOFORTE WIRES—ROLLED SASH BARS—COPPER-SHELLED ROLLERS—NOVEL EDGE TOOLS—FRAME-BLADED RAZORS—RACK ROLLERS.

THE year 1820 commences with a patent to William Hall and William Rostill, tortoiseshell box makers, for "certain improvements in the manufacture of hafts, handles, or hilts for knives, forks, swords, or any other instruments to which hafts, handles, or hilts are necessary, and can be applied, whether made of turtle or tortoise shell, or other suitable material." For some reason or other the patent never reached its final stage, so that no further information respecting the precise nature of the invention can be given beyond that which the title contains.—The name of Harcourt has long been known in connection with the cabinet brassfoundry trade, and several members of the family have contributed to the patent lists. In 1820 Mr. James Harcourt, struck by the enormous friction in the common castor, obtained protection for a method of lessening the friction by interposing small spheres or rollers between the moving parts, so as to allow the socket to turn on its axis with great ease. Being aware of the reproach brought against many contrivances of the sort—that they were rather more difficult to work than the ordinary combinations, by reason of the additional bearings—he proposed to confine the spheres or rollers in such a manner that no pivots were required, thus giving the utmost possible advantage.—It is not very often that the remains which have come down to us from prehistoric times can be made to furnish evidence of want of novelty of modern patents. Such, however, would appear to be the case with respect to that granted to Henry Botfield Thomason (son of Sir Edward), who, in 1820, patented a method of making cutlery, including knives of all sorts, scissors, razors, and surgical instruments, by soldering a narrow strip of steel into a groove in the edge of a blade made of gold or silver. The object was to produce articles having a rich appearance, and with a fine cutting edge ; but the idea happens to be very old indeed, for in the museum at Copenhagen there are knives made upon this very plan, which were found amongst the ancient Scandinavian remains so numerous in Sweden and Denmark. We are not aware of the precise age of these relics, but we believe that they have been used to support an argument that there was a period in the world's history when steel was a rare metal, and was accordingly used very sparingly.—The patent of William Kendrick, who is described as a chemist, for extracting tanning matter by macerating the materials in hot water under a pressure of from 8 to 12lbs. on the inch, forms one of the series of inventions which have done so much towards quickening the process of the manufacture of leather. Some say that rapidity has only been gained at the expense of quality. Thomas Dobbs's patent for plating tin upon lead was briefly referred to in a former article. It consists in pouring molten tin upon the sheet of lead to be coated, and rubbing it with a wisp of tow, a little rosin or turpentine being employed as a flux.

In the year 1821 we have to notice John Slater's cooking range, which was a rather complicated assemblage of contrivances for roasting, boiling, baking, and heating. The inventor, who was referred to in a previous article, seems to have enjoyed some amount of popularity, and to have done a considerable business at one time. We recollect seeing in the town one of the ranges made according to this specification about 35 years ago, but it found no favour with the presiding genius of the kitchen in which it was fixed.—The only improvement in William Lane's patent jack consisted in the use of two or more spring barrels instead of one, in order to give more power to the apparatus. By a suitable combination of wheelwork all the springs were caused to act upon the axis of the spit, and, as will be readily understood, the mechanism is of the simplest character.—The patent which was granted to Thomas Lees, the younger, snuffer maker, is one of a series of improvements in that article, the main object of which was to confine the snuff in a separate chamber which should remain closed when the jaws of the snuffers were separated for a fresh operation, and only open to receive the snuff just removed when it was fairly within the cavity of the snuffers. In that manner there was no possibility of dropping fragments of consumed wick during the act of snuffing the candle.—In the patent of Thomas Martin and Charles Grafton, printing ink makers, for making lamp black, we have a reference to a trade which is, we believe, no longer carried on in the town. They propose to call the product "spirit black," from the fact of "refined coal tar and coal tar spirits" being used in its manufacture. The substances are burnt in a specially-constructed furnace, the dense smoke produced being conducted through a succession of canvas bags where the suspended carbon is deposited, and from which it is shaken out occasionally.

We come now to the first patent taken out by Mr. Westley Richards, whose name is so closely connected with the gun trade of this town. The patent has reference to a condition of things now entirely passed away, but it is interesting as a land-mark in the history of invention, and as showing the form in which the "percussion system" was used sixty years ago. The idea of exploding the charge of a gun by the *detonation* instead of the *ignition* of the priming, was due to the Rev. Alexander Forsyth, a Scotch minister, who took out a patent in the year 1807. Forsyth's lock was in the main a contrivance by which a small portion of detonating composition was transferred before each shot from a magazine into a special chamber, where it was exploded by a pin struck by the hammer of the gun. The distributing mechanism was not altogether unlike that of an ordinary agricultural seed drill, which as is well-known consists of a hopper, at the bottom of which revolves a small cylinder having a series of receptacles sunk in it. As the cylinder rotates it transfers and allows to escape as much seed as is contained in one of these small hollows. As we have said, Forsyth supplied the detonating priming somewhat on the same principle. Westley Richards's primer was like an ordinary hammer in shape, fixed on a pivot between the nipple and the muzzle. The head was made hollow to contain a sufficient supply of composition for several shots, and when the gun was loaded the primer was let gently down on the nipple, where it remained until the piece was fired. It will be seen that before the hammer could take proper effect the primer must be moved out of the way, and this was done by a lever, one end of which was connected with the hammer, and the other pressed against the primer. When the trigger was pulled and the hammer released, the effect was to instantaneonsly throw back the primer, which left a small portion of detonating composition on the top of the nipple in readiness to be exploded by the blow of the hammer. Imbued, as we are, with deep respect for the labours of the *bona fide* inventor, we hesitate to speak of their labours as "clumsy," but it can easily be understood why the percussion system failed for many years to displace the old flint gun. Even that exquisitely simple contrivance, the percussion cap, was a long time in making its way, but more from imperfection in manufacture than from any inherent fault of principle. Mr. Richards subsequently took out patents for "primers," somewhat in the nature of caps, which were very highly spoken of by the sporting authorities of the day.

Only three patents were granted to Birmingham in the next year (1822), the first being to Benjamin Cook, the well-known tube maker, for a method of rendering fabrics and wood fire-proof, and for preserving wood from dry rot by the use of a solution of alkali. He mentions a variety of articles to which the composition may be applied, such as curtains, scenery of theatres, sail cloth, &c., and although the invention is sound enough in theory, it probably shared the same neglect as that which scores of other equally efficacious plans have met with.—William Brunton's patent smoke-consuming furnace was, we believe, the original progenitor of a large family of contrivances, in which the bars, and with them the fuel, were made to travel slowly from end to end of the furnace. Francis Deakin's method of making holster cases, cartouch boxes, dressing cases, &c., does not differ very much from that protected by his patent of 1812, described a few weeks ago. He uses sheet metal bent into shape in any suitable manner, which may be covered with calico, cloth, paper, leather, felted material, &c., and then varnished and lacquered.

As regards the patents of the following year (1823), we have to record Francis Deakin's improved method of stringing pianos, which permits the use of much harder wire than that usually employed. Instead of bending the string round a peg—which is not possible with hardened wire—the end is held in a screw clamp, the jaws of which are made slightly conical. The tuning pegs are also dispensed with, the strings being brought to the proper tension by means of a screw and nut.—Mr. Thomas Horne, the author of several ingenious inventions in the brass-foundry line, patented, in conjunction with William Bailey, of High Holborn, London, a method of finishing sash bars and metal mouldings by drawing. This was an improvement upon a process of rolling sash bars patented by Bailey in 1819. Instead of using an ordinary die for finishing the articles, they were drawn through rollers having suitable grooves cut in them. A superior article, the inventor says, is produced by means of these "roller dies" as they may be called, the use of which is a prominent feature in many of the pipe-making patents described in former articles. We may remark that the descendants of Mr. Horne's co-patentee still carry on the business of ironmongers in High Holborn.—Reference was made in the eighteenth article of this series to some of the plans which had been patented in the town for the purpose of obtaining copper shells for calico-printing rollers, and we have now to notice Thomas Attwood's patent, which consisted in soldering the shell to the mandril. It was not actually his own invention, having been communicated to him from abroad.—We are inclined to think that the patent granted to Bennington Gill, merchant, for an improvement in edge tools and cutlery, which had been communicated to him "by a certain foreigner residing abroad" was not *bona fide*, and it was certainly not new. In order to avoid the difficulty of tempering cutting instruments which have a very thin edge and a thick back, he proposed to make the blade of a plate of steel which was stiffened by a back made in the following manner. A piece of metal of the requisite thickness was turned up into the shape of a tube, but the edges did not quite meet so as to allow of the blade being inserted. The latter was firmly fixed in position by running in melted solder. Amongst the articles to be made in this manner the patentee mentions razors, choppers, saws, straw and hay knives, and scythes. The foreigner residing abroad was in all probability M. Boullay, cutler to the Royal Veterinary School at Alfort, who obtained a French patent for this invention in the same year. The original idea of these "frame-bladed razors," as they were called, was due to David Hartley, at one time M.P. for Hull. Articles made on this plan were manufactured in Paris, and it was there that Boullay saw them.—James Sprigg, the elder, fender maker, took out a patent for ornamenting fire-grates, fenders and fire-iron rests by the method known as "close plating;" rolled silver, cut to the required shape, being soldered upon iron or steel previously tinned.—Another of the inventions of Mr. Thomas Horne (mentioned above), was a method of manufacturing rack pulleys, the indentations in which were produced by rolling from a flat piece of sheet metal. The groves at the sides for guiding the carriage and pulley were formed by turning up the metal on a mandril, the entire rack being made of a single piece.

## CHAPTER XXXVI.

### MISCELLANEOUS INVENTIONS 1824-25.

SPIRAL SPRINGS FOR BELTS—PRINTING IMPROVEMENT —FRAMING FOR GLASS—CRIMPING MACHINE—CAST-IRON CONVERSION—FEATHERING PADDLE WHEEL—JOSEPH AND JAMES LUCKCOCK—"PILING" IRON BARS—MELTAL COVERED KNOBS—CHURCH'S SPIRAL CUT DRILLS—CORK-PACKED TAPS—CASTINGS BY COMPRESSED AIR—CALICO ROLLERS—HENRY JAMES'S RAILWAY PATENT—DIVING DRESS—SALT IN GAS MAKING—REVOLVING BOILER—"RIDING BITS" FOR SHIPS—"METAL-BOARD" BOOKS FOR SAFETY AGAINST FIRE.

THE use of a series of small spiral springs, arranged side by side and stitched between two layers of fabric for making elastic surgical belts, seems so obvious that one is almost inclined to think that it must have been known before 1824. However that may be, it was claimed as a new invention in that year by Henry Adcock, of Summer Hill Terrace, gilt-toy manufacturer. He states that springs made in that way are also suitable for being applied to the backs of waistcoats, trouser bands, &c., and he also describes two simple machines for coiling the wire of which the springs are composed.—That indefatigable inventor, Dr. Church, was again to the fore with an improved printing machine, which included a method of fixing the form on the table, a plan for ensuring a correction "register," mechanism for holding and taking off the sheets, and a mode of regulating the supply of ink.—The patent granted to James Wright Richards, of Caroline Street, metallic hot-house maker, for "an improved metallic frame and lap, applicable to hot-houses, greenhouses, horticultural frames and glasses, skylights, and other inclined lights and glasses," consists of a peculiarly shaped bar for forming the joints of the panes, the putty being confined in grooves and protected from the action of the atmosphere and from frost.—We come next to the "crimping, pleating, and goffering" machine invented by John Turner, brass and iron founder, which is simply a metallic box containing a heater with a fluted top of brass. The articles to be crimped are laid upon this box and a fluted roller traversed backwards and forwards until the required effect is produced.—Amongst the lesser lights who have been conspicuous in the Birmingham firmament, the two Luckcocks, Joseph and James, occupied no inconsiderable position. The former appears here by right of his patent for the use of common salt during the conversion of cast-iron into wrought in the puddling furnace. When the charge is just beginning to melt, a quantity of the above-named material is thrown in to act as a flux, and to improve the quality of the product. There was no novelty in the idea of employing salt during the reduction of iron, and it had been patented twice before, though not at the puddling stage, the process not having been invented at the time. This patent suggests the remark that in spite of the proximity of Birmingham to the South Staffordshire iron district, the number of patents for improvements in the manufacture of that metal—the actual smelting that is—which have originated in the town are remarkably few. In conjunction with Joseph Broomfield, engineer, Luckcock took out another patent in the year 1825, for a "feathering" paddle wheel. At a very early stage in the history of steam navigation it was perceived that much loss of power ensued consequent on the mode in which floats of a paddle wheel entered and left the water. To make this quite clear, let us consider the case of a wheel having only one float, placed radially according to the usual practice. As the wheel revolves the float will strike the water with its flat surface, which demands the expenditure of considerable engine-power, very little of which is effective in driving the vessel forward, and the same thing happens when the float is leaving the water. The wheel is in the most advantageous position for exerting its power when the float is in a perpendicular position, that is, when it is immersed to the greatest depth. In the feathering paddle wheel the floats are caused to turn on an axis, so that they enter and leave the water edgeways, and thus offer very little resistance. Luckcock and Broomfield proposed to accomplish this object by fixing a toothed wheel on the paddle shaft, which wheel gave motion in its turn to other wheels, and so caused the floats to revolve on their own axes. The invention consisted, in fact, of a peculiar application of the mechanical combination, known as the "epicyclic train." Another claim under the patent was for a rotatory engine—a machine which hundreds of inventors have tried their hands upon, but hitherto without practical success. The plan was ingenious, but it could not be made intelligible without drawings.

Joseph Luckcock was essentially a many-sided man, and perhaps some account of his career may not be unacceptable. He was born in the town on the 15th of January, 1760, his father having been engaged in business as a manufacturer of chemicals and metal refiner. At a very early age Joseph evinced a decided taste for science, especially the higher branches of the mathematics, and a few years later, when he was bound apprentice to his father, he commenced the study of chemistry. But his father's business proved a failure, and the son was apprenticed a second time to a buckle maker, and in due course he started on his own account in partnership with his brother. The decline of that business, in consequence of the change of fashion, left him after twelve years' work just where he began, and for a long period he occupied himself with concerns connected with the metallic arts. About the year 1817, finding that he had realised a modest competence, he built the well known round cottage at the corner of Wheeley's Hill Road and St. James's Road, Edgbaston, which his friends, in allusion to his astronomical studies, called Orbit Lodge. He published several papers on his favourite pursuit, including one on the theory of the tides, and another in *Newton's London Journal*, 1825, on the calculation of the solar year. In 1824, there appeared in the same journal an essay on the subject of light and heat. Some years before his death, finding that his income was hardly sufficient for his wants, he sold Orbit Lodge for a life annuity, and devoted himself during the remainder of his days to the study of physical science. He is said to have borne an extraordinary resemblance to the great Cuvier, whose portrait, prefixed to the memoir by Mrs. Lee, would serve as an excellent likeness of himself, making due allowance for the disparity of age. His brother James is known in the literary world by a series of lectures on "Moral Culture," published in 1817. He was also a correspondent of the *Monthly Magazine*, and in June, 1823, he contributed a remarkable article to that periodical, on the probable duration of our coal fields, a subject which has since attracted considerable attention. An interesting extract from a manuscript volume, written by James Luckcock, appeared in *Local Notes and Queries*, No. 626, published in July last. He contributed a very lengthy memoir of his brother Joseph to the *Mechanics' Magazine* of April 11, 1835.

Benjamine Cook, who appears to have been attacked with a mania for taking out patents on a great variety of subjects just about this time, specified a gun-lock with a spiral spring, either in place of the ordinary main spring or applied in a totally different manner.—The patent which was granted to Richard Hooton, of the Aqueduct Iron Works, iron manufacturer, is a puzzle. The specification is commendably short, and is as follows: "My invention consists in rolling iron from any number of pieces of puddled billets (more than one), which I fasten together by bands or otherwise." It is inconceivable that the method of "piling" bars was unknown until the year 1824, and it is equally difficult to understand how, if it was known, any iron manufacturer should have been so ignorant of what had been done as to spend a couple of hundred pounds or so in patenting an old process.—Benjamin Ager Day gives two separate specifications of his patent for making door knobs ; for the foundation of which he uses cast iron, and covers them with sheet brass in a lathe. He also claims the fixing of a rim of metal by the same means round wood knobs, and also a method of fixing the shanks in such knobs.—The year 1824 closes with an interesting invention of Dr. Church's for making what are known as "twist drills" and augers by winding a steel rod of peculiar section round a spirally-grooved mandril, a machine for effecting the operation being shown in the drawings. The form of the auger is claimed, whether it be produced as above described, or by cutting a spiral out of a solid bar.

In 1825 we have to chronicle William Rudder's patent for lining the barrels of cocks with cork, thus dispensing with the operation known as "tighting,"—that is, grinding the plug so as to exactly fit the barrel and prevent the escape of liquid. This was an improvement upon his patent of 1796, according to which the cork packing was not continuous but disposed in grooves cut round the plug, whilst here a complete cylinder or "bush" of cork was employed.—Dr. Church's improvement in casting rolls, pipes, &c., consisted in exhausting the air from the mould, and thus causing the molten metal to flow in ; the operation being assisted by the action of compressed air forced into the receptacle containing the metal. It will be, of course, understood that proper precautions must be taken to keep the mould and crucible quite air-tight. The inventor states that much sounder castings can be obtained by this method, which, by the way, has been reintroduced of late years.—In Thomas Attwood's method of fixing calico-printing rollers upon their mandrils, the ordinary slot and projecting "neb" are dispensed with, their purpose being served by notches at each end of the roller and corresponding projections upon the mandril. The mode of producing them is very ingenious, but it will take too long to describe it.—In the article upon road carriages, which appeared some few months back, we mentioned William Henry James, a clever engineer, long resident in the town. In 1825 he obtained a patent for a method of giving motion to railway trains, according to which every wheel throughout the train was *driven*, instead of, as now, merely forming a support for the carriages. This was accomplished by a shaft extending the entire length of the train, a sort of flexible coupling being provided between each carriage, so as not to interfere with the action of the mechanism when passing round curves. To give further facilities the rails at those parts of the line are made with several ribs or elevators the wheels being made to correspond by their peripheries being grooved, so that a wheel in effect possesses as many different diameters as there are variations in the surface of its periphery. By this means it may be made to travel faster or slower as may be desired. This is precisely the same *principle* as that now universally employed for compensating for the differing velocities of the inside and outside wheels of a railway carriage when passing round a curve, but the mode in which it is carried out is different, and, it must be a ded, very much simpler. —No particulars are extant of Badams's patent for obtaining metals from their ores, as no specification was filed.—The principle of W. H. James's diving apparatus consisted in the provision of a supply of compressed air contained in a sort of metal jacket which surrounds the upper part of the diver's body. Suitable valves were provided to reduce the pressure of the air before it was supplied to the diver. There were many schemes of a similar kind, but they were all superseded by what is known as the "close diving dress," which was the invention of Siebe.—The claim in the patent granted to J. F. Ledsam and Benjamin Cook is for the use of common salt in gasmaking, in whatever form or at whatever stage of the process of manufacture. For instance, it may be mixed with the coal in the retort, or used in solution as a purifying agent, through which the gas is passed to remove sulphur and other impurities. A very little knowledge of chemistry would have prevented the patentees—one of whom was an exceedingly clever man in his line—from wasting a large sum of money on a worthless patent.—The patent granted in this year to W. H. James, for an improved boiler has been made the foundation of a claim on his behalf to the invention of the ordinary "tubular boiler." An examination of the specification shows that there is not the slightest justification for such claim. He proposed to build up a boiler of a number of flat hollow rings which communicated with each other by means of a hole. The firegrate was in the interior, and the general appearance of such a boiler would be that of an ordinary Cornish boiler, with a rather large flue. He also proposed to mount it on a horizontal axis, and turn it round occasionally, so that each part should be exposed to the fire in succession. He hoped in that way to prevent the formation of scale.—The year closes with two patents to Benjamin Cook, one for the application of springs in any and every form to the "riding bits" of ships. In this way, when a heavy sea struck the ship when riding at anchor, the spring connection would enable the cable to "give" a little, and thus prevent it from snapping under the sudden strain. His other patent was for binding books in metal "boards," and making portfolios in the same way, so as to protect them from destruction or injury by fire. He thinks that his plan would be very useful in the case of registers and account books ; but we fear that his experience of the effect of fire upon books could not have been so great as is, unfortunately, that of many of the bibliophiles of the town. The covers would, of course, soon become red hot, in consequence of the high conducting power of the metal of which they are composed, and the fate of a pile of sheets of paper under such circumstances is tolerably evident.

## CHAPTER XXXVII.

### MISCELLANEOUS INVENTIONS.—1826-1830.

STAMPED IRON HANDLES—RISING HINGES—ECONOMICAL FILES—MECHANICAL STOKING—PLATED AND RINGED COINAGE—PURIFIED GAS—WATCHMAN'S TELL-TAIL—WOVEN WIRE SACKING—PATENT FITTED TUBES—PATENT CASTORS—STAMPED PICTURE FRAMES—LEAF TURN-OVER—WEAVING DOUBLE-FABRIC—SAFETY HARNESS—CRANK-AXLE COACH—KNIFE SHARPENERS—GLASS LUSTRES—GIRTHS AND SADDLES—CORRUGATED METAL—CONTINUOUS RAILS—SMOKE CONSUMING—SAFETY GUN STOCK—CONCLUSION.

THE patents of Samuel Whitfield, for producing handles, ears, and other mountings for culinary utensils, by stamping, have already been noticed in their proper place, and we commence the year 1826 with a further improvement in the same direction by William Whitfield—a relative, no doubt, of the Samuel formerly mentioned. The patent has reference to handles for saucepans, tea-kettle handle straps, and so forth, and, instead of forging them, the inventor forms what are technically called "moulds," or rough pieces of iron of such shape and thickness as would be sufficient to make the article. The piece of metal thus roughly shaped is heated to nearly a white heat, and subjected to the action of successive pairs of dies, the scale or dross being removed at each operation. The "fin" or web caused by the dies may be removed by filing, or by a cutting press. It is evident from this patent that the art of stamping objects in iron was only practised on a comparatively small scale fifty or sixty years ago.—The next patent is the product of the fertile brain of Benjamin Cook, who was particularly busy with inventions just about this time. The object is to produce what is known as a "rising" hinge, in which the opening in the joint and the disagreeable appearance of the wire in it are obviated by concealing the mechanism within a tube. (The earliest patent in which rising hinges are mentioned was granted in 1799 to one William King, a London tinplate-worker; but it is not quite clear that he was the first inventor of that ingenious device.) The same indefatigable inventor also took out a patent for an improved file, consisting of a steel strip cut with teeth, and fastened in any convenient manner on a bar or handle. The handles may be of hard wood or metal, as desired; and the inventor states that considerable economy would result from the adoption of his plan, much less steel being required.—The sash-fastener of John Loach consisted of a latch, something like that used for doors, which takes into notches in the upper sash or in the side of the frame. Several modifications are described, the leading idea being "to fasten two sashes, or other perpendicular sliding articles, with one action," and at any height.—It is difficult to see what novelty there could have been in the invention which was patented by Benjamin Lowe, gilt toy manufacturer, for pins with ornamental heads, or with artificial gem settings. No particular process of manufacture is claimed.—James Barron, brassfounder, patentee of an apparatus for feeding fuel to furnaces, was probably an importation from London, several patents having been granted to a person of the same name and business, residing in Wells Street, Oxford Street. The fuel is contained in a series of chambers one above the other, each having a hinged bottom secured by a catch in connection with a lever. When the catch is withdrawn the fuel contained in the chamber is shot into the trough leading to the furnace, and this operation is performed upon each chamber in succession at regular intervals by the descending weight of a clock. There is also an automatic arrangement set in motion by the weight of the falling fuel which actuates a rake and stirs the ashes out of the fire whenever more coal is put on. The rake is pivoted on an arm underneath the bars, and when the last box is emptied a bell rings to warn the attendant.—Dr. Church's specifications are, as a rule, too diffuse, and the inventions which they describe too complicated to permit of any useful abstract being given within reasonable limits. His patent for printing machinery granted in this year forms no exception, and all we can say is that it relates principally to improvements upon a former patent taken out in 1824.—The next invention we have to notice is, we fear, rather "Brummagem" in character, although it proceeded from a man whose name is usually associated with sound honest work. About the period of which we are treating there arose a demand for plated coinage, but the difficulty was to plate the edges and so prevent the base metal from being exposed. Sir Edward Thomason got over this difficulty by surrounding the "blanks" with a fillet, or band of solid metal, which was bent round and soldered firmly to the plated disc in the centre. The mode of manufacture as given in the inventor's *Memoirs* differs somewhat from that in the specification. He says in his book "from sheets of thin rolled pure silver I cut circles of the dimensions of the coin—say the dollar; this circle was by a lathe neatly turned over; it was then put into the collar of the dies; the force of the press was applied; the blow brought up the impression of the dollar, and the concussion fastened the silver to the edge of the dollar, and at the same time milled the edge. It was impossible at sight to tell these dollars from the real silver ones; the specific gravity, however, would detect them, also the sound of the ring." He was in treaty to execute a large number for a South American State, but the order was never carried out.

In the year 1827 we have a patent by J. F. Ledsam for purifying gas by passing it through a solution of chloride of ammonium, to which a quantity of lime is added. Sulphate of ammonium is occasionally employed intead of the chloride. In order to remove the unpleasant odour of coal gas the inventor proposes to expose it to the action of hypochlorite of sodium or calcium, either in solution or in a dry state.—Henry Knight, a clock maker of the town, took out a patent for a watchman's tell-tale, consisting of a clock dial with arrangements for making a mark upon it by pulling a lever from the outside of the case. If a permanent record of the watchman's vigilance is required, the marks are to be made on a piece of paper attached to the dial, and which is to be renewed every day, but otherwise they may be made on the dial itself, and effaced the next morning. With suitable modifications the apparatus may be used as an engine counter, as a chronograph, or for any purpose where intervals of time require to be measured and recorded.—Dr. Church's attachment for spindles of spinning machinery to prevent breakage of the threads, relates to a branch of manufacture of little interest here.—Thomas Breidenbach's patent for bedsteads consists in the use of worked or woven wire for sackings, testers, hangings, and coverings for excluding mosquitoes and noxious animals. For some purposes a folding screen covered with wire work may be extended round the bed.—The mode of action of B. J. Coombs's blind racks and blind rollers could not be made intelligible without the aid of drawings. It forms one of a long series of endeavours to produce a simple and effective method of holding the blind at any height.—The novelty in R. W. Winfield's patent tubes consisted in making them with a wood filling either solid or in segments, or with a hole from end to end. These tubes or rods are used in the construction of bedsteads, curtain and cornice poles, balusters, &c. The drawings annexed to the specification show several forms of tube, and the methods of connecting them together for various purposes.—The year 1827 closes with a patent granted to H. F. B. Geithner for castors. The improvement is small though important, and consists in making the pivot or pin with a circular shoulder or segment of a ball, which works in a corresponding cup or segment of a circle sunk in the top of the horn or claw. A much better bearing is ensured in this way, and friction is avoided.

In 1828, B. A. Day patented a mode of making metal picture frames, japanned or lacquered, to imitate those made of papiér mâche. They were produced by stamping in dies either in one piece or in lengths, in which latter case they were joined together by soldering. He makes his frames to represent "Gothic or Grecian architecture," which, if they were like the fire screens and chimney ornaments of the same inventor, referred to three weeks ago, must have been ugly enough.—The *volti subito*, or music leaf turner, patented by Abraham Louis, of Dean Street, mechanic, possesses the usual complications of these ingenious, but useless, contrivances. They will, no doubt, continue to be invented until the end of time, and musicians will probably be equally persistent in declining to make use of them.—In John Slater's axle box the rubbing part of the axle is covered by a brass box, having holes for admitting oil, and also grooves in the axle for distributing the oil to the parts where it is required.—James Simistor's patent for weaving double fabrics to be used in the manufacture of stays must have been a valuable one, inasmuch as it was thought worth while to incur the expense of a "disclaimer and memorandum of alteration," when the patent had very nearly expired, in order to make it quite secure, a part of the patent not having been found "useful." The details of the invention are of no interest, but the object may be briefly stated as a mode of weaving a double fabric with spaces left open between the thicknesses for the insertion of whalebone, or for cotton wool or other filling.

The year 1829 produced a patent by William Leeson, taken out in the name of his deceased partner, William Taft, for spring stop hooks for harness, which afforded greater facilities for attaching and detaching the various parts than the appliances then in use. In the event of the horse falling, the traces became immediately disconnected from the vehicle. The name is not very common, and it occurs to us to suggest that the Dr. Henry Beaumont Leeson who took out a patent for electro-plating, and who was connected with Messrs. Elkington, may possibly have belonged to the same family.—Thomas Brown, coach maker, appears as the inventor of an improved coach, in which the load was brought as low as possible by the use of a cranked axle, the luggage being stowed in a capacious "boot" in the hind part of the vehicle. There were also other minor improvements.—Dr. Church patents a marine boiler and a device for feeding fuel to furnaces. He also patented a knife sharpener, consisting of a pair of conical discs, cut with file-teeth, which by a very happy idea, he proposes to attach to a carving fork, just where the ferrule is usually placed. The sharpener would thus always be ready for use, whilst it would not interfere with the use of the fork. He also shows in the drawings a larger instrument, on the same principle, suitable for sharpening scythes. The discs are made by stamping pieces of steel with a serrated die.—No particulars can be given of the "self-acting air or gas regulator, or stop-cock," patented by George Danre, as the patent never arrived at maturity.—Thomas Osler's patent for chandeliers relates to a method of combining prismatic or other shaped pieces of glass, so as to produce ornamental and refractive forms for chandeliers or other articles used in ornamental lighting. A variety of methods are shown, but the combinations produced have long been out of fashion, though we believe that the same plan, or a modification of it, is still used in building up the large lighthouse lenses, first suggested by Fresnel, modified by Stevenson, and developed by Messrs. Chance Brothers.

The inventions of the year 1830 comprise a patent for an improved spring girth for saddles, granted to John Lawrence, silversmith, of this town, and William Rudder, of Edge, Gloucestershire. The principle claimed is the application of springs to saddle-girths, so that they shall always be tight, notwithstanding any temporary change in the size of the animal.—In the same year Thomas Thatcher, saddler, patented an elastic saddle, having the property of adapting itself to the particular shape of the horse upon which it may be placed.—Dr. Church patented the use of corrugated metal in the construction of ships, carriage bodies, gas-holders, water tanks, distillers' and brewers' vessels, &c., so as to give greater strength with the same weight of metal. In the case of ships the air enclosed between the inner planking and the corrugated sides would serve to increase the buoyancy. With reference to the claim for carriages, it may be pointed out that Dr. Church's plan was applied to railway trucks many years afterwards.—Allusion has been made in a former article to the attention which Birmingham men have devoted to the subject of calico-printing rollers, and Benjamin Cook's patent of this year furnishes another instance of it. It consists in forming the "web" or projection on the copper shell—which takes into a corresponding slot in the mandril—by cutting away the superfluous metal. No machinery is, however, described by which the result is accomplished.—An American merchant, one Augustus Whiting Gillet, patented as a communication from George Bridgman, the oft-repeated device of making the wheels of a carriage travel on the inside of the rim of a larger wheel or hoop, and thus form a continuous rail, which is, as it were, lain down and taken up again directly the wheels of the vehicle have passed over. This has been patented many times in various forms, but the present inventor was not the first to describe it.—Dr. Church's patent for steam boiler and other furnaces consisted mainly in forcing a descending blast into the fuel, for the purpose of increasing the combustion and consuming the smoke.—A contribution to the numerous safety gun locks which has been proposed was made by Bartholomew Redfern, who patented a mode of discharging a gun by pressing a stud situated on the upper part of the stock close to the breach of the gun. Under ordinary circumstances the stud is covered up and protected from accidental pressure. A separate and distinct operation is necessary to expose the stud, and put it in a position for being acted upon to discharge the pieces.

The industrial history of the town has now been brought down to the end of the year 1830, and in the case of the more important trades which have been dealt with in special articles, to quite recent times. It will have been noticed that many of the miscellaneous inventions are either unconnected with the manufactures of the district, or are in themselves of no great importance. In fact, the town, so far as its inventive ingenuity was concerned, begins to be more cosmopolitan, and therefore not so interesting for our present purpose. We do not propose to pursue the subject further, and accordingly break off, leaving undescribed the miscellaneous inventions of the last 50 years. as well as those industries which have arisen since the year 1830. In relative importance of the town as regards the number of new inventions brought forward was discussed in the opening article of this series, so that there is no occasion to repeat what was there said. Those who have followed us from week to week will have no difficulty in acknowledging that the mechanical combinations, and the new processes which have originated in Birmingham are neither few nor insignificant. What is perhaps of more importance, it was here that many new modes of manufacture took their rise, and which are now of almost universal application in the production of an innumerable variety of articles.

[THE END.]

# APPENDIX.

## List of Patents granted in Birmingham from the earliest Period down to the Passing of the Patent Law Amendment Act, October 1st, 1852.

[In the following List the titles of the patents have been shortened or altered, so as better to express the precise nature of the invention. Where the full address of the patentee is given, it has been retained, for purposes of identification; but, in the majority of instances, the inventor is described as "of Birmingham, in the County of Warwick," or "of the Parish of Aston, in the County of Warwick." In cases of joint patents, where some of the patentees do not belong to the town, the place of residence of the "foreign" patentee is given in full. The inventor's business or calling has also been retained (except when he is simply described as a "manufacturer"), as serving at times to prove the existence of particular trades as separate branches of manufacture.]

### 1722.

Richard Baddeley, Ironmonger, "Casting wheel streaks and box irons."

### 1738.

Lewis Paul, Gentleman, "Spinning cotton."

### 1742.

John Baskerville, Japanner, "Metal plates and mouldings."

### 1748.

Lewis Paul, Gentleman, "Carding engine."

### 1759.

Stephen Bedford, Japanner, "Ornamenting surfaces by transferring impressions of engravings."

Thomas Blockley, Locksmith, "Polishing and rolling iron, and making wheel tyres."

### 1762.

JOHN ROEBUCK, Doctor of Physic, "Making malleable iron."

### 1768.

SAMUEL and SAMPSON FREETH, Merchants, "Hand grinding mills."

GEORGE WHATELEY, Silver Plater, "Plating and drawing wire."

GEORGE WHATELEY, Silver Plater, "Plating and drawing wire."

### 1769.

RICHARD FORD, Smith, "Rolling, drawing, and stamping metals."

JOSEPH ASHTON, "Casting nails."

### 1770.

JOHN SMITH, Jeweller, "Buttons."

MAYER OPPENHEIM, Merchant, "Red glass."

### 1771.

JOSEPH ASHTON, Caster, "Casting nails."

### 1772.

WILLIAM TUTIN, Buckle-maker, "Blueing buckles."

HENRY CLAY, Japanner, "Papier mâché."

### 1773.

WILLIAM SMALL, Doctor of Physic, "Clocks."

### 1774.

ISAAC WHITEHOUSE, Enameller, "Imitation enamelling."

### 1775.

JOHN IZON and THOMAS WHITEHURST, Brassfounders, "Hinges."

### 1776.

RICHARD DEARMAN, Ironmaster, "Plantation hoes."

### 1777.

JOHN MARSTON, Brassfounder, and SAMUEL BELLAMY, Engraver and Diesinker, "Stamping metals."

### 1778.

HENRY CLAY, Japanner, "Papier mâché buttons."

WILLIAM COLLINS, Button and Toy Maker, "Plated metal for buttons, &c."

JOHN TALBOT, Filesmith, "Machine for grinding metal bars."

## 1779.

Richard Dearman, Ironmaster, "Metal mills for grinding grain."

Jonathan Taylor, Ironfounder, "Cast iron hollow-ware."

John Keir, Gentleman, "Alloy of iron, zinc, and copper."

William Bell, Jeweller and Toymaker, "Rolling buttons, buckles, toys, &c."

## 1780.

James Watt, Engineer, "Copying letters."

Thomas Lindopp, Brassfounder and Factor, "Saddles."

William Bache, Chemist, "Motive power."

James Pickard, Button Maker, "Crank for Steam Engines."

Daniel Winwood, Ironmaster, "Buckle chapes."

## 1781.

John Waters, Gun and Pistol Maker, "Pistol with a bayonet."

Daniel Winwood, Buckle Chape Maker, "Buckles."

James Watt, Engineer, "Steam Engines."

William Playfair, Handsworth, Engineer, "Rolling spoons, forks, medals, &c."

## 1782.

James Watt, Engineer, "Steam Engines."

## 1783.

William Collins, Manufacturer, "Ships' bolts."

John Westwood, Button Maker, "Rolling copper and other metal rods."

## 1784.

James Watt, Engineer, "Steam engines."

## 1785.

John Cooper, Japanner, "Japanning, or blueing buckles."

James Watt, Engineer, "Furnaces."

James Alston, Buckle and Button Maker, "Plating buckles, &c."

## 1786.

John Eginton, Engraver, "Wood buttons."

John Francis, Founder, "Cast iron buttons."

Joseph Antley, Spur Maker, "Spurs."

Thomas Cheston, Plater, "Spring Buckles."

Edward Strickland, Mechanic, "Thief and fire alarum."

Joseph Ashton, Bordesley, Ironfounder, "Casting buttons and button shanks."

Henry Clay, Japanner, "Buttons."

Obadiah Westwood, Button Maker, "Papier mâché articles."

## 1787.

JOSEPH RABONE, Button Maker, "Bone and ivory buttons."

ISAAC WHITEHOUSE, Toy Maker, "Casting plates and dishes, and plates as substitutes for warming pans."

JOHN MILES, Gentleman, "Lamps."

JOHN ROSE, Button Maker, "Steel buttons, buckles, sword hilts, &c."

JAMES ALSTON, Button Maker, "Metal buttons."

JAMES YATES, Brassfounder, "Stamping metals."

## 1788.

JOSEPH JUKES, Plater, "Substitute for buckles."

## 1789.

WILLIAM SHEWARD, Needle Maker, "Needles."

## 1790.

RICHARD LAW, Plater, "Buckles."

SAMUEL HANDS, Buckle Maker, "Ornamenting buckles."

HENRY CLAY, Japanner, "Slate or stone buttons."

WILLIAM COLLINS, (of Lambeth), Engine Maker, and CHARLES WYATT, Merchant, "Tinned copper sheets."

JOHN WHITWORTH, Deritend, Diesinker, "Plating tin with silver, and making plated goods."

MATTHEW BOULTON, Engineer, "Coining press."

WILLIAM WHITMORE, Tool Maker, "Button shanks."

## 1791.

CHARLES WYATT, Merchant, "Pipes, cisterns, &c., of tinned copper."

## 1792.

WILLIAM WHITMORE, Engineer, "Mashing machine."

JAMES SMITH, Buckle Maker, "Spring fastening for shoes, spurs, &c."

HENRY CLAY, Japanner, "Panels for coaches, furniture, &c."

## 1794.

JOHN HAND, Plater, "Plated cutlery."

RALPH HEATON, Jobbing Smith, "Button shank machine."

BENJAMIN ADAMS, Plated Buckle Maker, "Buckles."

## 1795.

JAMES TIMMINS, Button Maker, "Beer cocks."

JOHN GREGORY HANCOCK, Artificer, "Sash bars, fencing &c."

JONATHAN GROVE, Manufacturer, "Gun locks."

## 1796.

William Whitmore, Engineer, "Weighing machines."
William Rudder, Cockfounder, "Cocks."
Henry Clay, Esquire, "Tilting mechanism for carts."
John Gregory Hancock, Diesinker, "Embossing paper."
Edward Thomason, Button Maker, "Self-acting carriage steps."

## 1797.

Matthew Boulton, Esquire, "Hydraulic Ram." *(Communicated by Montgolfier.)*

## 1798.

Henry Clay, Esquire, "Saving water in canal locks."
Francis Farquharson, Factor, "Brick-making machine."
Francis Hollick, Curry-Comb Maker, "Curry combs."
William Docker, Slate Merchant, "Stone pipes."
Philip Chell, Engineer, and Henry Nickholls, Carpenter, "Lift for canals."
Edward Thomason, Button Maker, "Self-acting carriage steps."

## 1799.

Samuel Sandy Hickling, Gentleman, "Enamelling hollow-ware."
Edward Thomason, Manufacturer, "Gun locks."

## 1800.

Joseph Barnett, Merchant, "Buttons."
Isaac Hadley Reddell, "Carriages to travel on land or water."
Isaac Hadley Reddell, "Stirrups."
Thomas Gill, "Rifling machine for fire-arms."
Isaac Hadley Reddell, "Carriages."

## 1801.

William Pritchard (of Shrewsbury), Hatter, and Thomas Willmore, Japanner, "Hats, helmets, &c."
Joseph Chirm, Pump Maker, "Boring wood pipes."
Josiah Longmore, Gun Lock Maker, "Pattens or clogs."

## 1802.

Abner Cowell Lea, Manufacturer, "Umbrella furniture."
Edward Thomason, Manufacturer, "Corkscrews."
John Barnett, Toy Maker, and Joseph Barnett (of Warwick), Cutler, "Parasols and umbrellas."

## 1803.

EDWARD THOMASON, Button and Toy Manufacturer, "Telescopic hearth brushes."

## 1804.

EDWARD THOMASON, Button and Toy Manufacturer, "Pikes."

THOMAS DOBBS (of King's Norton), Chemist, "Lead, plated with tin (Albion metal.)"

JOHN GREGORY HANCOCK, Die Engraver, "Presses and stamps."

RICHARD KENTISH, late Captain Cambridgeshire Militia, "Armour waistcoat."

## 1806.

JOHN JONES (the younger), Tool Maker and Diesinker, "Gun barrels."

THOMAS BOURNE and WILLIAM CHAMBERS, Scale Beam Makers, and CHESTER GOULD, Mechanic, "Roasting jack."

HENRY PRATT, Steel Toy Maker, "Toast Stand."

## 1807.

CHESTER GOULD, Gentleman, "Dial Weighing Machine."

EDWARD COKE WILMOT, Gentleman, "Warming pans."

## 1808.

EDWARD MOORE NOBLE, "Surgeon, "White lead."

JOSEPH JOHNSON and JOHN WILMOT, Manufacturers, "Warming pans."

BENJAMIN COOK, Manufacturer, "Gun barrels and ramrods."

WILLIAM BELL, Engineer, "Pipes or pumps for liquids."

JOSEPH WILLMORE, Silversmith, and JOHN TONKS, Plater, "Making nails."

JOSEPH MASON GUEST, Thread Manufacturer, "Making thread."

EDWARD THOMASON, Manufacturer, "Umbrellas and parasols."

WILLIAM TOMPSON, Deritend, Locksmith, "Locks."

## 1809.

SAMUEL WHITFIELD, Church Street, Brazier and Scale Beam Maker, "Stamping ears and handles for culinary utensils, &c."

PHYLLIS BOWN THOMASON, wife of Edward Thomason, Manufacturer, "Umbrellas and parasols."

THOMAS WELLS, Erdington, Cockfounder, "Cocks."

## 1809 — (*continued*).

CHARLES NORTON, Builder, "Use of cast iron in buildings."

JOHN JONES, Gun Lock and Barrel Maker, "Gun barrel skelps."

EDWARD GRIFFITH, Optician, "Air-tight agitable lamp."

## 1810.

WILLIAM MURDOCK, Engineer, "Boring stone pipes and fashioning stone columns."

THOMAS BAYLEY, Brassfounder, "Blind racks."

JOHN SLATER, Coach Spring Maker, "Hanging grindstones."

WILLIAM DOUGHTY, Engineer, "Gaining mechanical power by an epicyclic train."

JOHN BOSWORTH, Coal Dealer, "Discharging loads from carts."

WILLIAM BELL, Handsworth, Engineer, "Circular shears for cutting pasteboard, leather, straw, &c."

WILLIAM SHAKESPEAR and THOMAS OSLER, Glass Chandelier Manufacturers, "Glass drops for chandeliers."

BENJAMIN AGER DAY, Handsworth, Brassfounder, "Toast stand and combined hearth brush and toasting fork."

SAMUEL HOBDAY, Woodcock Street, Aston, Snuffer Maker, "Snuffers."

## 1811.

ABRAHAM WILLIS, Deritend, "Steel toys, including barbers' curling irons, sugar nippers, snuffers, &c."

ROBERT DAVIS, Umbrella and Parasol Furniture Manufacturer, "Umbrella and parasol furniture."

HENRY JAMES, "Propelling boats on canals."

WILLIAM FINCH, Ironmaster, "Nails."

WILLIAM JENKINS, Brassfounder, "Handles for cabinet work."

THOMAS ATTWOOD, Esquire, and BENJAMIN COOK, Gilt Toy Maker, "Brass-covered tubing, rods, cornice poles, &c."

RALPH SUTTON, Brassfounder, "Blind racks."

HENRY JAMES, Merchant, and JOHN JONES, Gun Barrel Maker, "Gun barrels."

JOHN STUBBS JORDEN, Patent Copper Window Frame Manufacturer, "Glazing hothouses."

WILLIAM WALKER JENKINS, Brass Founder, "Drawer knobs."

WILLIAM RUDDER, Cock Founder and Warming Pan Maker, "Cocks."

### 1812.

George Babb, Bordesley, Engineer, "Files, edge-tools, and other steel articles."

Philip Chell, Engineer, "Rotary pump and rotary steam engine."

John Loach, Brassfounder, "Castors, knobs, and lock furniture."

William Whitfield, Steelyard Maker, "Steelyards."

Bassett Burrows, Hatter, "Waterproof hats."

John Scambler, Needle Manufacturer, "Needles."

John Simpson, Tin-plate Worker, "Lamps."

Henry Osborn, Bordesley, "Iron tubes and gun barrels."

Francis Deakin, Deritend Mills, Wire Drawer, "Sheaths and cases for knives, &c."

Benjamin Cook, Gilt Toy Maker, "Applications of brass-cased tubing."

Henry Osborn, Bordesley, Sword and Gun Barrel Maker, "Welding gun barrels and tubes."

### 1813.

Thomas Ryland, Plater, "Fenders."

Charles Plimley, "Rolling file blanks and similar articles."

James Timmins, Manufacturer of Sashes and Hothouse Lights with metal bars, "Hothouses and horticultural buildings."

Robert Lewis, Brassfounder, "Metal chimney pieces."

Samuel Whitfield, Brazier, "Mountings for culinary and similar utensils."

William Jenkins, Brassfounder, "Socket castors."

Edward Thomason, Manufacturer, "Whips."

Henry Osborn, Whitmore House, "Gun barrels."

Edward Biggs, Brassfounder, "Working a stamp by power."

Ralph Sutton, Brassfounder, "Safety gun lock."

### 1814.

John Slater, Manufacturer, "Washing and drying apparatus."

Emanuel Heaton, Gun Finisher, "Waterproof pan for gun locks."

William Whitfield, Scale Beam Maker, "Carriages."

William Neville, Merchant and Coachfounder, "Hurdles, gates, palisades, espalier frames, &c."

John Stubbs Jorden, Copper Sash Maker, "Horticultural buildings."

James Dobbs, Gentleman, "Reaping machine."

1815.

THOMAS BAGOT, Surveyor, "Lift for canal boats."

WILLIAM BELL, Engineer, "Rolling wire."

1816.

G. F. MUNTZ, Metal Roller, "Smoke consuming furnace."

HENRY OSBORN, Bordesley, "Welding tubes."

EDWARD BIGGS, Brassfounder, "Making pans and stails."

JOHN ASTON WILKES, Glass Toy Maker, "Glass icicles, spangles, &c."

JOHN PIERCY, Tortoiseshell Toy Maker, "Thimbles."

1817.

HENRY OSBORN, Bordesley, "Sword and Gun Barrel Maker, "Welding Tubes."

BENJAMIN COOK, Gilt Toy Maker, "Calico printing rollers."

BENJAMIN AGER DAY, "Chimney ornaments."

THOMAS TAFT, Saddler, Whip Maker, and Bridle Cutter, "Bridle reins and bits."

EDWARD BIGGS, Brassfounder, "Stamping pans and handles for the same."

JOHN TURNER, Button Maker, "Plating copper or brass with gold."

1818.

JOHN SIMPSON, Plater, "Spring hooks for harness."

THOMAS JONES, Bradford Street, Iron Founder, and CHARLES PLIMLEY, Refiner, "Blowing engines."

SAMUEL HOBDAY, Snuffer Maker, "Snuffers."

WILLIAM FINCH, Gentleman, "Hollow bits for horses."

HENRY PERSHOUSE, Factor, "Stamping scale pans."

1819.

JOHN SIMPSON, Plater, "Harness."

SAMUEL HAYCRAFT, Spoon Maker, "Rolling and stamping spoons."

WILLIAM NEALE, Whitesmith, "Motive power."

JOSEPH WEATHERLEY PHIPSON, "Metal Merchant, "Metal tubes."

JAMES MASON, Merchant. "Method of working oars." *(A Communication)*.

STEPHEN BEDFORD, Birchall Street, Iron Founder, "Manufac ture of iron."

WILLIAM BRUNTON, Engineer, "Revolving furnace grate and steam engine piston."

JOHN HEARD, Joiner, "Cooking apparatus."

## 1820.

WILLIAM HALL and WILLIAM ROSTILL, Tortoiseshell Box Makers, "Knife and other handles."

WILLIAM BRUNTON, Engineer, "Furnaces with revolving grate."

WILLIAM KENDRICK, Chemist, "Tanning."

JAMES HARCOURT, Brassfounder, "Castors."

HENRY BOTFIELD THOMASON, (son of Edward Thomason), "Cutlery."

WILLIAM KENDRICK, Chemist, "Tanning."

THOMAS DOBBS, Smallbrook Street, Plater, "Plating lead with tin."

## 1821.

JOHN SLATER, "Kitchen range."

WILLIAM LANE, Jack Maker, "Roasting jack."

THOMAS LEES, the younger, Snuffer Maker, "Snuffers."

THOMAS MARTIN and CHARLES GRAFTON, Printing Ink Makers, "Lamp black."

SAMUEL HOBDAY, Patent Snuffer Maker, "Umbrella furniture."

WILLIAM WESTLEY RICHARDS, Gun Maker, "Gun lock."

## 1822.

BENJAMIN COOK, Patent Tube Maker, "Preserving and fire-proofing wood."

WILLIAM BRUNTON, Engineer, "Furnace with travelling grate."

FRANCIS DEAKIN, Sword Maker and Wire Drawer, "Holster cases, cartouch boxes, &c."

## 1823.

FRANCIS DEAKIN, Sword Maker, "Stringing pianofortes."

WILLIAM BAILEY, Ironmonger, of High Holborn, London, and THOMAS HORNE, the younger, of Belmont Row, Brass-founder, "Drawing sash bars and mouldings."

FRANCIS DEAKIN, Wire Worker, "Umbrella furniture."

THOMAS ATTWOOD, Banker, "Calico-printing rollers." *(A Communication).*

BENNINGTON GILL, Merchant, "Cutlery and edge tools." *(A Communication).*

JAMES SPRIGG, the elder, Fender Maker, "Grates, fenders, and fire-iron rests."

THOMAS HORNE, the younger, Brassfounder, "Blind racks."

## 1824.

HENRY ADCOCK, Summer Hill Terrace, Gilt Toy Maker, "Spring belts and bandages."

WILLIAM CHURCH, "Printing machinery."

JAMES WRIGHT RICHARDS, Caroline Street, Metallic Hothouse Maker, "Skylights, hothouses, &c."

JOHN TURNER, Brass and Ironfounder, "Crimping, plaiting, and goffering machine."

WILLIAM CHURCH, "Casting metals."

JOSEPH LUCKCOCK, Round Cottage, Edgbaston, Gentleman, "Manufacture of iron."

WILLIAM HENRY JAMES, Coburg Place, Winson Green, Engineer, "Steam carriages."

JAMES COOK, Gun Maker, "Gun locks."

RICHARD HOOTON, Aqueduct Ironworks, Iron Manufacturer, "Piling iron bars for rolling."

BENJAMIN AGER DAY, Fire Screen Maker, "Drawer and door knobs."

WILLIAM CHURCH, "Augers and boring bits."

JAMES and WILLIAM HENRY DEYKIN, Button Makers, "Buttons."

## 1825.

WILLIAM RUDDER, Edgbaston, Cockfounder, "Cocks."

WILLIAM CHURCH, "Casting metals under pressure."

THOMAS ATTWOOD, Manufacturer, "Calico printing rollers."

WILLIAM HENRY JAMES, Coburg Place, Winson Green, Engineer, "Railways and carriages therefor."

AARON JENNENS and JOHN BETTERIDGE, Japanners in Ordinary to his Majesty, "Ornamenting papier mâché with pearl shell."

JOHN BROOMFIELD, Islington, Engineer, and JOSEPH LUCKCOCK, Gentleman, "Feathering paddle wheel and rotary steam engine."

JOHN BADAMS, Ashted, Chemist, "Smelting metals."

WILLIAM HENRY JAMES, Coburg Place, Winson Green, Engineer, "Diving apparatus."

JOSEPH FREDERICK LEDSAM, Merchant, and BENJAMIN COOK, Brassfounder, "Purifying gas by the use of common salt and metallic salts."

WILLIAM HENRY JAMES, Coburg Place, Winson Green, Engineer, "Steam boilers."

BENJAMIN COOK, Brassfounder, "Riding bits for ships."

BENJAMIN COOK, Brassfounder, "Binding books."

## 1826.

WILLIAM WHITFIELD, "Handles for saucepans, &c."

BENJAMIN COOK, Brassfounder, "Rising hinges."

BENJAMIN COOK, Brassfounder, "Files."

JOHN LOACH, Brassfounder, "Sash fastener."

BENJAMIN LOWE, Gilt Toy Maker, "Ornamental pins."

JOHN BARRON, Brassfounder and Venetian Blind Maker, "Feeding fuel to furnaces."

WILLIAM CHURCH, "Printing machinery."

EDWARD THOMASON, Goldsmith and Silversmith, "Coining medals, tokens, &c., of plated metal."

JOHN GREGORY HANCOCK, Plated beading and canister hinge Manufacturer, "Umbrellas."

## 1827.

JOSEPH FREDERICK LEDSAM, Merchant, "Purifying coal gas."

THOMAS BREIDENBACH, Merchant, "Bedsteads."

HENRY KNIGHT, Clock Maker, "Watchman's tell-tale."

WILLIAM CHURCH, "Spinning machinery."

THOMAS BREIDENBACH, Merchant, "Bedsteads."

BENJAMIN MERRIMAN COMBS, Ironmonger, "Blind racks."

ROBERT WALTER WINFIELD, Brassfounder, "Tubes for bedsteads, &c."

THOMAS TYNDALL, "Buttons" (*Communicated by Dr. Church*).

DANIEL LEDSAM and WILLIAM JONES, "Nails."

FREDERICK BENJAMIN GEITHNER, Brassfounder, "Castors."

THOMAS TYNDALL, "Nails" (*Communicated by Dr. Church*).

## 1828.

THOMAS BREIDENBACH, Merchant, "Tubes."

BENJAMIN AGER DAY, Brassfounder, "Picture frames."

ABRAHAM LOUIS, Dean Street, Mechanic, "Music leaf turner."

JOHN SLATER, Maker of Coach Springs and Axletrees, "Axles and boxes."

JAMES SIMISTER, Bull Street, "Fabric for stays."

## 1829.

WILLIAM TAFT, Harness Maker, Saddler, and Bridle-cutter, "Harness."

WILLIAM CHURCH, Bordesley Green, "Covered buttons."

BENJAMIN COOK, Brassfounder, "Calico-printing rollers."

WILLIAM LEESON, "Harness" (*Communicated by Wm. Taft, deceased*).

THOMAS BROWN, Coachmaker, "Coaches."

1829 — (*continued*).

WILLIAM CHURCH, Heywood House, "Marine engines, paddle wheels, boilers, furnaces, &c."

WILLIAM CHURCH, Heywood House, "Knife sharpener."

GEORGE DANRE, "Gas regulator."

THOMAS OSLER, Chandelier Furniture Maker, "Glass chandeliers."

1830.

ENOCH WILLIAM RUDDER and ROBERT MARTINEAU, Cock-founders, "Cocks."

JOHN LAWRENCE, Silversmith, and WILLIAM RUDDER (of Edge, in the County of Gloucester, Gentleman), "Spring girths for saddles."

THOMAS THATCHER, Saddler, "Elastic saddles."

WILLIAM CHURCH, Heywood House, Bordesley Green, "Using corrugated iron in building ships, carriages, gasometers, &c."

JOHN, WILLIAM, GEORGE, and REUBEN HEATON, "Steam carriage."

BENJAMIN COOK, Brassfounder, "Calico-printing rollers."

AUGUSTUS WHITING GILLET, Merchant, "Carriage wheels." (*Communicated by George Bridgman*).

WILLIAM CHURCH, Heywood House, "Furnaces for steam boilers, evaporating pans, &c."

BARTHOLOMEW REDFERN, Gunmaker, "Safety gun lock."

1831.

WILLIAM WESTLEY RICHARDS, Gunmaker, "Touch hole and primer."

JOHN GUNBY, George Street, Sand Pits, Artist, "Combining glass and metal for ornamental purposes."

SAMUEL HOBDAY, Steel Snuffer and Toy Maker, "Rotary steam engine."

RICHARD PROSSER, Civil Engineer, "Bullion nails."

WILLIAM CHURCH, Heywood House, Bordesley Green, "Nail-making machinery."

AUGUSTUS WHITING GILLET, Merchant, "Metronome." (*A Communication*).

BENJAMIN AINGWORTH, Button Maker, "Buttons."

JOSEPH GILLOTT, Steel Pen Maker, "Pens."

DANIEL LEDSAM, Manufacturer, and WILLIAM JONES, Screw Manufacturer, "Pins, rivets, screws and nails."

ROBERT WALTER WINFIELD, Brassfounder, "Bedsteads, &c."

## 1832.

William Church, Bordesley Green, "Steam carriages."

William Church, Bordesley Green, "Nail-making machine."

John Day, Brassfounder, "Cocks."

Benjamin Cook, Brassfounder, "Utilizing the alcoholic vapours given off from fermenting dough for the manufacture of varnish and for other purposes."

George Edmonds, St. Mary's Square, Gentleman, "Philosophic alphabet."

Benjamin Cook, Brassfounder, "Applications of zinc to the manufacture of various articles."

Harriot Grant Gillet (widow of Augustus Whiting Gillet, merchant), "Metronome."

John Jacob Parker (late of Sheffield), Gentleman, "Fountain pens."

Thomas Wells Ingram, Die Sinker, "Press for making horn buttons."

Alexander, and William Southwood Stocker, Union Rolling Mills, "Iron shoe tips and heels, chain links, &c.

George Frederick Muntz, Metal Roller, "Muntz's metal."

George Frederick Muntz, Metal Roller, "Ships bolts of Muntz's metal."

## 1833.

Thomas Moore Evans, Merchant, "Dressing flax and hemp." (*A Communication*).

William Harrold, Merchant, "Paper making machinery." (*A Communication*).

Edward Lucas, Edward Street, "Pumps."

Charles Jones, Gun Maker, "Gun locks."

Joshua Horton, Taylor's Dock, Boiler Maker, "Chains."

William Shilton, Machinist, "File-cutting machine."

John Holmes, Engineer, "Metal button shanks."

Charles Jones, Gun Maker, "Gun locks."

William Church, Heywood House, Bordesley Green, "Steam engines and boilers."

James Windeyer Lewty, Lichfield Street, Brassfounder, "Castors."

George Frederick Muntz, Metal Roller, "Application of Muntz's metal in the manufacture of boilers."

Daniel Ledsam and William Jones, Screw Manufacturers, "Machine for making pins and needles."

## 1834.

THOMAS GRIFFITHS, Tin Plate Worker, "Making tea kettles, &c."

GEORGE MILLICHAP, Carriage Axle Maker, "Locomotive carriages."

RICHARD WALKER, "Metal gun wads."

JOHN GOLD, Glass Cutter, "Cutting glass articles."

JOHN ASTON, Button Maker, "Covered buttons."

GEORGE JOSEPH GREEN, JOHN OGDEN BACCHUS, and WILLIAM GAMMON, Glass Makers, "Glass making." *(A Communication).*

GEORGE RICHARDS ELKINGTON, Optician, "Spectacles."

MANOAH BOWER and GEORGE BLYTH, Merchant, "Saddles."

## 1835.

BENJAMIN, and JOSEPH COOK, Brassfounders, "Beds and mattresses."

SAMUEL BURRELL, Gilt Toy Maker, "Buttons."

RICHARD PROSSER, Civil Engineer, "Nail making machine."

WILLIAM CHURCH, Heywood House, Bordesley Green, "Marine engines, locomotives, steam carriages, railways, &c."

RICHARD HILL, Merchant, "Locks and staples."

CONRAD GEORGE KUPPLER, of Nuremberg, at the Polytechnical Institution, but now of Birmingham, "Weighing machines."

THOMAS HORNE, Brassfounder, "Hinges."

EDWARD JONES, Builder and Brick Maker, "Brick-making machine."

WILLIAM BANKS, Spring Hill Terrace, "Ruling and pressing paper."

WILLIAM LUCY, Miller, "Steam Engines."

THEODORE SCHWARTZ, Bradford Street (formerly of Stockholm), Technologist, "Engine worked by compressed gas or vapour."

JOHN BIRD, Gentleman, "Printer's ink and other pigments."

HENRY ADCOCK, Summer Hill Terrace, Engineer, "Working cranes by tidal power."

HUMPHREY JEFFERIES, Goldsmith and Jeweller, "Buttons."

JOHN SAMUEL DAWES, Ironmaster, "Use of hydrogen gas in the blast furnace."

ROBERT GRIFFITHS, Machine Maker, "Rivets, screw-blanks, and bolts."

## 1836.

James Barron, Brassfounder, and Edward Thomas, workman to James Barron, "Bedsteads and accessories."

Anthony Theophilus Merry, Metal Dealer, "Plated German silver."

William Westley Richards, Gun Maker, "Percussion caps."

William Brindley, Caroline Street, Paper Maker, "Papier mâché trays."

John Jeremiah Rubery, Umbrella and Parasol Furniture Manufacturer, "Umbrella and parasol stretchers."

John Holmes, Engineer, "Steam boilers."

William Taylor, Smethwick, and Henry Davies, of Stoke Prior, Worcestershire, "Boiler feeding apparatus, rotary steam engines, and propelling mechanism for ships."

Henry Elkington, Gentleman, "Rotary steam engine."

Manoah Bower, "Hoods or heads of carriages."

John Woolrich, Professor of Chemistry at the Royal Medical School, "Carbonate of baryta."

George Richards Elkington, Gilt Toy Maker, "Gilding metals."

Alexander Stocker, Gentleman, "File-cutting machinery."

William Southwood Stocker, Machinist, "Nails, rivets, and screw blanks."

Robert Griffiths, Machine Maker, and John Gold, Glass-cutter, "Grinding glass and stone spangles and drops."

Henry Van Wart, Gentleman, and Samuel Aspinwall Goddard, Merchant, "Steam carriages and steam engines."

Frederick Benjamin Geithner, Brassfounder, "Ball bearings for blinds, maps, looking-glasses, &c."

Alexander Stocker, Bordesley Ironworks, and Henry Downing, French Walls Ironworks, "Rivets, screw-blanks, &c."

## 1837.

Robert Griffiths, Smethwick, Machine Maker, and Samuel Evers (of Cradley Ironworks), Iron Manufacturer, "Nuts."

Henry Adcock, Summer Hill Terrace, Engineer, "Smelting furnaces."

James Cook, Gun Maker, "Gas burners."

Henry Elkington, Gentleman, "Gilding and platinizing metals."

## 1837 — (*continued*).

HENRY ELKINGTON, Gentleman, "Steam engines, boilers, and furnaces."

HENRY CHRISTOPHER WINDLE (of Walsall), Merchant, JOSEPH GILLOTT, Metallic Pen Maker, and STEPHEN MORRIS, Artisan, "Steel pens and penholders."

THOMAS WELLS INGRAM, Horn Button Maker, "Horn Buttons."

JOHN LOACH, Brassfounder, "Blind furniture."

JOHN JEREMIAH RUBERY, "Umbrella furniture." (*A Communication*).

ELISHA WYLDE, Engineer, "Locomotive and other engines."

HENRY ELKINGTON, Gentleman, "Gilding and silvering metals."

JEREMIAH BYNNER, Lamp Maker, "Deflectors for gas lamps."

BENJAMIN COOK, Brassfounder, "Argand gas burners."

WILLIAM ELLIOTT, Button Maker, "Covered buttons."

WILLIAM HENRY JAMES, Civil Engineer, "Telegraphs."

WILLIAM BRINDLEY, Patent Paper Tray Maker, "Screw presses."

## 1838.

WILLIAM WESTLEY RICHARDS, Gun Maker, "Primer for fire arms."

ROBERT LUCAS CHANCE, Glass Works, Smethwick, "Glass making." (*A Communication*).

JOSEPH ROCK COOPER, Gun Maker, "Fire arms."

JAMES TIMMINS CHANCE, Glass Maker, "Grinding, polishing, and flattening sheet glass."

JOHN RATCLIFF, Lamp Maker, "Lamps."

ROBERT MARTINEAU and BROOK SMITH, Cock Founders, "Cocks and valves."

THOMAS WALKER, Clock Maker, "Rotatory steam engines."

CHARLES GREEN, Gold Plater, "Brass and copper tubes."

EDWARD WHITE BENSON, Manufacturing Chemist, "White lead."

GEORGE ROUND, Lock Filer, and SAMUEL WHITFORD, Die-sinker, "Manufacturing parts of gun-locks by stamping."

HENRY ELKINGTON, Gentleman, "Steam and air engines."

HENRY VAN WART and SAMUEL ASPINWALL GODDARD, Merchants, "Railway signals, coupling for railway carriages, and valves, &c., for locomotives."

## 1838 — (*continued*).

JOB CUTLER, Ladypool Lane, Gentleman, and THOMAS GREGORY HANCOCK, Mechanist, "Steam engine condensers, and boiler-supply apparatus."

GEORGE RICHARDS ELKINGTON and OGLETHORPE WAKELIN BARRATT, "Coating and colouring metals."

JOSEPH RAYNER, JOSEPH WHITEHEAD RAYNER (both of Birmingham), and HENRY SAMUEL RAYNER (of Ripley), Civil Engineers, "Spinning fibrous materials."

HENRY GIBBS, Button Maker, "Metal buttons."

ALEXANDER SOUTHWOOD STOCKER and CLEMENT HEELEY, "Trouser straps."

JOHN WOOLRICH, Professor of Chemistry, "White lead."

JOHN FOWLER, Gentleman, "Sulphuric acid."

LUKE HEBERT, Bristol Road, Civil Engineer, "Embalming and preserving corpses." *(A Communication.)*

THOMAS MAYOS WOODYATT, Screw Manufacturer (of Cookley, in the County of Worcester), and SAMUEL HARRISON, Mechanic, "Making screws."

JOB CUTLER, Gentleman, Ladypool Lane, "Metals and tubes made therefrom."

## 1839.

OGLETHORPE WAKELIN BARRATT, Metal Gilder, "Soda making."

JOHN EVANS, Paper Maker, "Air pumps for paper-making machines."

RICHARD PROSSER, Civil Engineer, "Boilers, stoves, &c."

JAMES RUSSELL, Handsworth, Gas Tube Manufacturer, "Welded iron tubes." *(Prolongation of Whitehouse's Patent of 1825.)*

JOB CUTLER, Gentleman, Ladypool Lane, "Chains for suspension bridges, &c."

THOMAS HENRY RYLAND, Screw Maker, "Screws."

GEORGE STOCKER and JOSEPH BENTLEY, Gun Makers, "Guns and pistols."

JOSEPH GILLOTT, Steel-pen Maker, and THOMAS WALKER, Machinist, "Safety-valves; railway carriage buffers; wheels and axles for carriages; spark arresters; substitute for the crank."

JAMES BARLOW, Brassfounder, "Candlesticks." *(A Communication.)*

### 1839 — (*continued*).

RICHARD PROSSER, Civil Engineer, "Nail and Screw-making machinery."

WILLIAM HARPER (of Cooper's Court, Cornhill, London), Patent Stove Manufacturer, and THOMAS WALKER, Machinist, "Stoves and Grates."

GEORGE STOCKER, Brassfounder, "Cocks."

JOHN HENRY RODGERS, Merchant, "Clasps, buckles, hooks and eyes, &c." (*A communication.*)

LUKE HEBERT, Civil Engineer, "Lamps." (*A Communication*).

CHARLES OSBORNE, Cork-screw Manufacturer, "Cork-screws."

ALEXANDER SOUTHWOOD STOCKER, Union Rolling Mills, and THOMAS JOHNSON (of Ridgacre Ironworks, Staffordshire), "Iron tips and heels for shoes."

JOB CUTLER, Ladypool Lane, Gentleman, "Alloys."

DAVID HARCOURT, Brassfounder, "Castors."

HENRY VENNER COCKS, Ironfounder, "Stoves and furnaces."

LUKE HEBERT, Patent Agent, "Packing presses." (*A Communication.*)

JOHN EVANS, Paper Maker, "Preparing and cleansing paper machine felts."

JAMES GUEST, Machinist, "Locks and other fastenings." (*A Communication.*)

WILLIAM NEWMAN, Brassfounder, "Blind furniture."

THOMAS HARDEMAN CLARKE, Cabinet Maker, "Fasteners for sashes, tables, windows, &c."

### 1840.

JOSEPH ROCK COOPER, Gun Maker, "Fire arms and balls for the same."

CHARLES ROWLEY, Stamper and Piercer, and BENJAMIN WAKEFIELD, Bordesley, Machinist, "Cutting-out, stamping, and piercing buttons, washers, &c."

JOB CUTLER, Ladypool Lane, Gentleman, and THOMAS GREGORY HANCOCK, Highgate, Machinist, "Cutting corks and forming necks of bottles."

WILLIAM BRINDLEY, Northwood Street. "Baling presses."

JOSEPH SHORE, Merchant, "Protecting metallic surfaces by the deposition of copper or nickel."

LUKE HEBERT, Civil Engineer, "Spades and shovels."

HENRY SMITH, Lamp Maker, "Gas burners and lamps."

GEORGE RICHARDS ELKINGTON and HENRY ELKINGTON, "Coating metals."

1840 — (*continued*).

Alexander Southwood Stocker, "Gas and other tubing."

Richard Prosser, Cherry Street, Civil Engineer, "Welded tubes."

John Gold, Etna Glass Works, "Moulding decanters, finger plates, and chandelier drops."

William Potts, Brassfounder, "Suspending pictures and curtains."

Henry Holland, Darwin Street, Umbrella Furniture Maker, "Umbrellas and parasols."

Rice Harris, Gentleman, "Use of glass as a printing surface for calicoes, &c. ; calico-printing rollers."

William Southwood Stocker, "Nails, pins, and rivets."

Christopher Dain, Edgbaston, Gentleman, "Inkstands."

Alexander Southwood Stocker, "Tubes for gas, &c."

Richard Prosser, Civil Engineer, and John James Rippon (of Wells Street, Oxford Street, London), Ironmonger, "Stoves and cooking apparatus."

Richard Prosser, Civil Engineer, "Making buttons, knobs, rings, tiles, &c., by compressing clay in fine powder."

Thomas Barnabas Daft, Gentleman, "Inkstands."

Theophilus Richards, Merchant, "Sawing veneers."

Luke Hebert, Civil Engineer, "Polishing the eyes of needles."

William Church, Civil Engineer, "Hooks and eyes."

Thomas Horne, Brassfounder, "Hinges."

John Whitehouse, the younger, Birchall Street, Brassfounder, "Spring hinges and door springs."

George Goodman, Needle and Pin Maker, "Mourning and dress pins."

Alexander Dean and Evan Evans, Millwrights, "Mills with a grinding surface of porcelain, and dressing machines."

Frederick Payne Mackelcan, Civil Engineer, "Thrashing machine and horse gear."

Josiah Pumphrey, New Town Row, Brassfounder, "Hooks and eyes."

Charles Winterton Baylis, Clerk, "Pens and penholders."

Elias Robison Handcock, late of the 9th Regiment, "Turntables, castors, axleboxes and bearings."

1840 — (*continued*).

FREDERICK PAYNE MACKELCAN and JAMES MURDOCH (of Hackney Road, London), Civil Engineers, "Tables and other furniture and castors." *(Partly a Communication).*

JOSEPH PARKES, Button Manufacturer, "Covered buttons."

1841.

WILLIAM LACEY, Agent, "Ornamenting Glass for various purposes; knife handles; window frames, and ships lights."

JOHN LOACH, Brass Founder, "Castors."

THOMAS GRIFFITHS, Tin Plate Worker, "Dish covers."

CHARLES GREEN, Gold Plater, "Brass and copper tubes."

ANTHONY THEOPHILUS MERRY, Metal Refiner, "Furnace for smelting zinc, lead, &c."

ROBERT WALTER WINFIELD, "Metallic bedsteads and castors."

ALEXANDER PARKES, Artist, "Electro-depositing works of art."

HENRY McEVOY, Graham Street, Hook and Eye Maker, "Trouser straps and fastenings for wearing apparel."

WILLIAM LITTELL TIZARD, Brewer, "Brewing apparatus."

HUMPHREY JEFFERIES, Button Maker, "Buttons."

ALEXANDER SOUTHWOOD STOCKER and CLEMENT HEELEY, "Patten and clog-ties, and hooks and eyes."

PHILEMON AUGUSTINE MORLEY, "Sugar moulds, dish covers, &c."

JOSEPH BETTRIDGE, Wood Turner, "Pillars for lamps, &c."

JOSEPH RATCLIFF, "Hinges."

WILLIAM WALKER JENKINS (of Greet, Yardley), "Making and papering pins."

THOMAS CHAMBERS and FRANCIS MARK FRANKLIN (both of Laurence Lane, London), Warehousemen, and CHARLES ROWLEY, Button Maker, "Buttons and dress fastenings."

LUKE HEBERT, Temple Row, "Making gas and gas burners."

OGLETHORPE WAKELIN BARRATT, Metal Gilder, "Electro-depositing metals."

LUKE HEBERT, "Fulling woollen cloth." (*A Communication*).

SAMUEL STOCKER (of Barford Street, London), Engineer, and GEORGE STOCKER, Cockfounder, "Pumps, water closets, cocks and pipe joints."

THOMAS WELLS INGRAM, Manufacturer, "Shears, snuffers, and rag-chopping machines. (*Partly a Communication*).

BENJAMIN AINGWORTH, Gentleman, "Buttons."

1841 — (*continued*).

JOHN JONES, Smethwick, Engineer, "Steam Engines."

HENRY DAVIES, Engineer, "Drilling and planing metals and cutting wheels."

JEREMIAH BYNNER, Lamp Maker, "Gas burners."

JOB CUTLER, Ladypool Lane, Gentleman, "Boiler tubes."

HENRY DAVIES, Engineer, "Propelling and drawing boats and carriages."

JOSIAH TAYLOR, Brassfounder, "Lamp for burning solid fatty matters."

ROBERT HENDERSON, China Dealer and Glass Stainer, "Heating and lighting apartments."

WILLIAM WESTLEY RICHARDS, Gun Maker, "Gun locks and percussion caps."

WILLIAM CHURCH, Civil Engineer, and JONATHAN HARLOW, "Tubes and bedsteads."

THOMAS STARKEY, Copper Cap Maker, "Percussion caps."

WILLIAM CARRON, Lathe Maker, "Clogs and pattens."

WILLIAM ROBINSON KETTLE, Waterloo Street, Accountant, BENJAMIN WAKEFIELD, Ryland Street North, Civil Engineer, and WILLIAM CROSHER, Cumberland Street, Screw Maker, "Expanding bolts for connecting wooden framework together."

1842.

JOHN JEREMIAH RUBERY, Umbrella and Parasol Furniture Manufacturer, "Umbrella and parasol furniture."

DANIEL GREENFIELD, the elder, Brassfounder, "Door knobs."

GEORGE CARTER HASELER, Jeweller and Toy Maker, "Tops of scent bottles."

HENRY BARRON RODWAY, Wine Merchant, "Horse shoes."

ZECHARIAH PARKES, "Domestic flour mill with dressing apparatus."

THOMAS CLIVE, Ironfounder, "Candlesticks."

SEPTIMUS COCKINGS, Draughtsman, "Lamps, candlesticks, snuffers, and extinguishers." (*Partly a Communication*).

JOSEPH GIBSON, "Axles and Axle-boxes."

BENJAMIN COOK, junior, Brassfounder, "Ornamenting bedsteads with papier mâché."

BENJAMIN AINGWORTH, Gentleman, "Plate-glass."

JOSEPH CHATWIN, Lamp Maker, "Cocks."

JAMES TIMMINS CHANCE, Glass Manufacturer, "Sheet-glass."

JOSEPH SCHLESINGER, "Inkstands and letter clips."

1842 — (*continued*).

ROBERT BENTON, Land Agent, "Propelling and retarding carriages on railways."

JOHN CHATWIN, Button Maker, "Covered buttons."

JOHN STEPHEN WOOLRICH, Chemist, "Use of sulphites for electro-plating, and applying magneto-electric currents for the same."

THOMAS WALKER, Stove Maker, "Stoves."

RICHARD FORD STURGES, "Producing a matted or frosted surface on Britannia metal and plated wares."

JAMES INSOLE, Saddlers' Ironmonger, "Brushes."

JOHN MITCHELL, Steel Pen Maker, "Pens and penholders."

CHARLES ROWLEY and JAMES TURNER, Button Makers, "Metal buttons."

CHARLES HEARD WILD, Engineer, "Fire-proof floors."

THOMAS MANSELL, Agent, "Cutting out and shaping boot fronts."

EBENEZER TIMMIS, "Extinguishing fires, and fire-escape."

CHARLES HEARD WILD, Engineer, "Railway switch."

JAMES BROWN, of the firm of Boulton, Watt, and Co., Soho, Engineers, "Marine engines."

WILLIAM BRIDGES, Button Tool Maker, "Buttons."

JOSEPH BEAMAN, Smethwick, Ironmaster, "Making malleable iron."

JOSEPH ROCK, jun., Factor, "Locks."

1843.

HENRY HUSSEY VIVIAN (of Singleton, Glamorganshire), and WILLIAM GOSSAGE, Manufacturing Chemist, "Furnaces for smelting zinc and other ores."

JOHN FREARSON, Machinist, "Hooks and eyes, fastenings for trouser straps, clogs, &c."

THOMAS SIMPSON, "Buckles."

FREDERICK COOK MATCHETT, "Hinges."

JOHN ASTON and WILLIAM ELLIOTT, Button Makers, "Covered buttons."

RICHARD PROSSER and JOB CUTLER, Civil Engineers, "Pipes and bars and applications of the same."

RICHARD FARMER, Upholder and Cabinet Maker, and JOSEPH PITT, Plumbers' Brass Founder, "Water closet, rotary pump, and method of tightening the sacking of bedsteads."

**1843** — (*continued*).

Thomas Wells Ingram, Engineer, "Ornamenting wood by pressing and embossing."

Oglethorpe Wakelin Barratt, Experimental chemist, "Gilding and plating metals."

William Needham, Gunsmith, "Gun lock and cap charger."

Alexander Parkes, Artist, "Obtaining and applying solutions of india-rubber."

Henry Clarke Ash, "Tea pots."

Henry Smith, "Door fastenings and alarums."

John Charlton, Factor, "Castors."

Jonathan Saunders, Soho Hill, Gentleman, "Wheel tyres and axles."

Thomas Morton Jones, Merchant, "Furnaces and vessels for heating liquids and gases."

Frederick Isaac Welch, "Producing a grained surface on enamelled leather."

John Withers, Smethwick, Manufacturing Manager, "Making glass."

James Roose, Gentleman, "Welding gun barrels and ordnance."

Joseph Rock, junr., Factor, "Locks and latches."

Alexander Southwood Stocker, Wine Merchant, "Stoppering bottles, and corkscrews."

Benjamin Cook, junr., Merchant, "Decorating various metallic articles with papier mâché."

**1844.**

John Hinks, George Wells, and Joseph Finnemore, Pen Makers, "Pens."

John Fielding Empson, "Buttons, hooks and eyes, and other dress fastenings."

William Lucas Sargant, "Gun barrels."

James Silcock, Engineer, "Planes."

Benjamin Aingworth, Gentleman, "Covered buttons."

Joseph Gibson, junior, Japanner, "Ornamenting glass."

Alexander Parkes, Artist, "Metallic alloys and depositing metals."

William Sheldon, Japan Painter, "Papier mâché buttons, trays, &c."

Frederick Stephenson, High Street, Comb Maker and General Dealer, "Bookbinding, and combined paper cutter and penholder."

1844 — (*continued*).

EMANUEL WHARTON, Engineer, "Metallic piston and packing for disc engines."

JAMES HARDY, Gentleman, "Welding tubes."

WILLIAM TAYLOR, Door Spring Maker, "Casting pulleys for sash lines and covering hat pegs with sheet brass."

ISAIAH DAVIES, Engineer, "Rotary steam engines."

JOHN LOACH, "Corkscrews, and cocks, taps and valves."

WILLIAM ELLIOTT, Button Maker, "Covered buttons."

THOMAS FARMER, "Ornamenting papier mâché and japanned goods."

CHARLES NOSSITER, Lyndon End, Tanner, "Tanning."

JAMES GEORGE NEWEY, Hook and Eye Maker, and JAMES NEWMAN, Jeweller, "Buttons, fastenings for trouser straps, and other dress fastenings."

WILLIAM BEDINGTON, jun., "Furnaces."

CHARLES HENRY CAPPER, Engineer, "Palisades, gates, and fences."

ALEXANDER PARKES, Artist, "Metallic alloys and depositing metals."

WILLIAM NEWMAN, Brassfounder, "Window blinds."

GEORGE MILLICHAP, "Axle-trees."

JOHN JEREMIAH RUBERY, Umbrella and Parasol Furniture Manufacturer, "Umbrellas and parasols."

1845.

ROBERT GRIFFITHS, Smethwick, Engineer, "Bolts, spikes and rivets."

EDWIN LUCAS, Machinist, "Chains."

THOMAS TURNER CHATWIN, Button Maker, and GEORGE SEYMOUR, Tool Maker, "Covered buttons."

GEORGE JOSEPH GREEN, "Rein-holder for saddles or carriages."

OGLETHORPE WAKELIN BARRETT, Experimental Chemist, "Condensing the gases evolved during various manufacturing operations."

GEORGE SELBY, Cambridge Street, Tube Maker, "Welded tubes."

STEPHEN PERRY (of Woodlands Place, St. John's Wood, London), Gentleman, and THOMAS BARNABAS DAFT, Manufacturer, "Springs for girths, belts and bandages, and making elastic bands."

JOHN LORD, Friday Bridge, Merchant, "Feeding steam boilers." (*A Communication*).

1845 — (*continued*).

RICHARD PROSSER, Civil Engineer, "Metal tubes."

CHARLES JAMES SMITH, Whittall Street, Gun Maker, "Cap charger for fire arms, and powder flasks."

CHARLES WILLIAM FIRCHILD,* Gentleman, "Machine for cutting, slicing, rasping, and grinding."

CORNELIUS WHITEHOUSE, Gun Barrel Maker, "Welding gun barrels and tubes."

WILLIAM LUCY, Miller and Baker, "Kneading machine."

JOHN READING, "Clasps and similar dress fastenings."

WILLIAM COSTON AITKEN, Clerk, "Ornamenting cornices, ends for cornice poles, curtain bands, &c., with glass, porcelain, &c."

JAMES HARDY, Gentleman, "Metal tubes."

THOMAS SILVER SHAW, Pawnbroker, "Roasting jacks."

EDMUND RATCLIFF, "Door lock furniture."

DALRYMPLE CRAWFORD (of Stratford-on-Avon †), "Dibbling machine."

GEORGE DANIEL BISHOPP, Edgbaston, Civil Engineer, "Rotatory engines."

ALEXANDER PARKES, Artist, "Coating and ornamenting metals."

THOMAS HOLLINGSWORTH, Cigar Merchant, "Cigar cases."

CHARLES NOSSITER, Lyndon End, "Leather."

WILLIAM ELLIOTT, Button Maker, "Buttons."

DALRYMPLE CRAWFORD, Gentleman, "Railway brakes."

THOMAS EDWARDS, Islington Foundry, Engineer, "Steam engines."

ISAAC HAWKER BEDFORD, "Glass."

WILLIAM HANNIS TAYLOR (of Piccadilly, London), Gentleman, and FRANCIS ROUBILIAC CONDER, Civil Engineer, "Atmospheric railways."

HENRY PERSHOUSE, Manufacturer, "Inkstands, pen wipers, and attaching postage stamps and wafers."

---

* Patents were granted in 1842 to "Charles William Firchild, of Weoley Park, Northfield, farmer," for propelling ships, and in 1844 to "Charles William Firchild, of Leamington, gentleman," for a rotary engine.

† Mr. Parkes in his report on the implements at the Newcastle meeting of the Royal Agricultural Society, in 1846, refers to this inventor as "of Birmingham."

## 1846.

THOMAS MOORCROFT BENBOW, Surgeon, "Surgical bandages, hooks and eyes, and stays."

EDWIN CHESSHIRE, Surgeon, "Safety buffer for obviating the effects of railway collisions."

WILLIAM GREENER, Gun Maker, and WILLIAM EDWARDS STAITE (of Peckham, London), Gentleman, "Electric lighting."

CHARLES ROWLEY, Button Maker, "Buttons."

THOMAS PEMBERTON, junior, "Introducing metal, glass, china, &c., into fringe used in upholstery; curtain bands, bell-pulls, &c., of metal in imitation of ropes and tassels."

JOHN BANFIELD, Organ Builder, "Pneumatic whistles for communicating between engine drivers and guards, or between the different stations on railways."

CHARLES ILES, Bordesley, Machinist, "Carding buttons, hooks and eyes, &c."

ALEXANDER PARKES, Artist, "Preparing gutta-percha and india-rubber."

EDWIN COTTERILL, "Door and window furniture."

PETER BISHOP, Bayonet Manufacturer, "Bayonets."

JAMES TIMMINS CHANCE, Glass Maker, and HENRY BADGER (of West Bromwich), Glass Stainer, "Glass making."

WILLIAM CHURCH, Civil Engineer, "Stamping and cutting metals; tubes, hollow rails, &c."

CHARLES THOMAS LUTWYCHE, Gold and Silver Chain Maker, "Porcelain buttons."

EDWARD ALFRED COWPER, Smethwick, Engineer, "Moulding railway chairs."

JOHN ASTON, Button Maker, "Buttons and dress ornaments."

THOMAS WALKER, Stove Maker, "Ships' logs and sounding apparatus."

WILLIAM MILL, Hunter's Lane, Handsworth, "Pencils, pens, penholders, and inkstands."

GEORGE DOWNING, Steel-pen Tool Maker, "Penholders."

WILLIAM MIDDLEMORE, "Saddles."

THOMAS PAYNE, Handsworth, Gentleman, "Rolls for rolling metals."

CHARLES ILES, Bordesley, Gentleman, "Presses for cutting and stamping."

1846 — (*continued*).

Daniel Sydney Hasluck, Merchant, "Saddles." (*A Communication*).

Alexander Parkes, Artist, "Candles."

Richard Ford Sturges, Britannia Ware Maker, "Filters, and mode of casting tea-pots."

Charles Fox, London Works, Smethwick, Engineer, "Shearing, cutting, and punching metals."

George Frederick Muntz, M.P., "Alloy for sheathing ships."

Arthur Millward, Gentleman, "Engraving and ornamenting metallic surfaces by the use of voltaic electricity."

David Barnett, Merchant, "Calculating machines."

Thomas Craddock, Engineer, "Steam engines and boilers."

John Todd, Glasgow, and William Johnston, Engineers, "Arranging the rails in engine or railway carriage sheds, and an improved switch box."

1847.

Thomas Deakin, King's Norton, Engineer, "Machinery for cutting, stamping, and pressing metals."

Thomas Barnabas Daft, Gentleman, "Inkstands and fastenings for elastic bands."

John Loach, Brassfounder, "Window and other fastenings."

Charles Fox, London Works, Smethwick, "Press for welding and forging metals."

Amedee François Remond, "Steam engines and boilers."

Louis Nicolas de Meckenheim, Machinist, "Nails, screw-blanks, rivets, bolts, and pins."

Morris Lyons, Chemist, and William Millward, Silver Operator, "Metallic alloy and deposition of metals."

Samuel Kendrick, Handsworth, Ironfounder, "Moulding and casting various articles."

Theodore Hyla Jennens, "Ornamenting papier mâché and other articles."

Richard Peyton, and Jonathan Harlow, Metallic Bedstead Makers, and Thomas Horne, Brassfounder, Bordesley Works, "Tubes and bedsteads."

Henry McEvoy, Newhall Street, Machinist, "Hooks and eyes."

James Timmins Chance, Handsworth, Glass Maker, "Making glass."

John Hunt, Brassfounder, "Gas and other burners."

### 1847 — (*continued*).

Theodore Fletcher, Brassfounder, "Protecting the silvering of glass specula by a coating of copper, &c.

Arthur Boyle, Umbrella Frame Maker, "Metal buttons."

James Simister, "Stay busks and fabric for stays and belts."

Thomas Horne, "Carriage windows."

Brooke Smith and Richard Ford Sturges, "Filters."

George Heaton, Engineer, "Counterweights for locomotives."

Alexander Parkes, Experimental Chemist, "Extraction of metals, and coating iron and steel with copper and its alloys."

John Britten, Machinist, "Cooking ranges and apparatus, bread slicers, sugar nippers, &c.

Mary Jenkins, Aston, "Hooks and eyes, and mode of blueing various articles."

### 1848.

Job Cutler, Sparkbrook, Civil Engineer, and Charles Robinson, Gentleman, "Welded boiler tubes."

John Hickman, Clerk, "Connections for bedsteads, and attaching handles to doors, drawers, &c."

James Timmins Chance and Edward Chance, Glass Makers, "Glass furnaces and glass making."

Thomas Potts, Brass Tube Maker, "Boiler tubes."

Charles Green, Patent Brass Tube Maker, and James Newman, "Railway wheel tyres."

John Britten, Machinist, "Stoves, ventilating apparatus, gas burners, carriage windows, door springs, and fastenings."

Charles Fielding Palmer, Druggist, "Artificial chalybeate water."

Alexander Parkes, Experimental Chemist, "Extracting metals from their ores and coating metals."

Isaiah Davies, Engineer, "Rotary and other steam engines."

William Taylor, Mechanist, "Skelp-bending machine."

William Brindley, "Ornamenting papier mâché in relief, and making basons and hats of the same material."

Charles Henry Capper, Edgbaston, Gentleman, "Rotating sieve and jigger for preparing ores, &c."

Samuel Thornton, Merchant, and James Edward McConnell (of Wolverton), Engineer, "Locomotives, brakes, couplings and signals for railway trains."

Charles Rowley, Button Manufacturer, "Buttons."

## 1848 — (*continued*).

ROBERT WALTER WINFIELD and JOHN WARD, a workman in the employ of R. W. Winfield, "Drawing taper tubes and making covered tubes for gas-fitting."

JOHN FREARSON, Islington, Machinist, "Chains, nails and pins."

JOSEPH GILLOTT and JOHN MORRISON, "Ornamenting wooden penholders and similar articles."

WILLIAM CHURCH, Civil Engineer, and THOMAS LEWIS, Woollen Draper, "Cutting, perforating, and pressing card."

ROBERT WALTER WINFIELD, "Bedsteads."

CHARLES ILES, Machinist, "Carding hooks and eyes, buttons, pins and needles, and making up tapes, ribbons, &c., for sale."

ALEXANDER and HENRY PARKES, "Smelting metals and forming alloys."

HENRY NEWSON, Smethwick, Engineer, "Trusses."

WILLIAM BAKER, Civil Engineer, and JOHN RAMSBOTTOM (of Longsight, near Manchester), Engineer, "Railway wheels, turntables, and bearings for shafts."

## 1849.

WILLIAM CHAMBERS DAY, Ironfounder and Weighing Machine Maker, "Platform weighing machine."

RICHARD FORD STURGES, Britannia-ware Maker, "Casting candlesticks and lamp pillars."

AMEDEE FRANÇOIS REMOND, "Envelope-making machine."

JOB CUTLER, Sparkbrook, Civil Engineer, "Metal tubes."

THOMAS HENRY RUSSELL, Tube Maker (of Wednesbury), and JOHN STEPHEN WOOLRICH, Chemist, "Coating metals."

CHARLES WILLIAM SIEMENS, Engineer, "Steam engines, and evaporating liquids."

ALEXANDER PARKES, Harborne, Chemist, "Alloys, and depositing metals."

CHARLES GREEN, Patent Brass Tube Maker, and JAMES NEWMAN, "Railway wheels."

JOHN BRITTEN, "Cooking ranges, roasting jacks, &c."

SAMUEL ALFRED CARPENTER, "Substitute for buckles."

1849 — (*continued*).

CHARLES ILES, Machinist, "Combining coloured silk waste, &c., with plastic matters for making inkstands, picture frames, &c., and also with cement for ornamenting walls."

SAMUEL ALLPORT, Gun Maker, "Making weavers' mails."

WILLIAM GOOSE, "Making nails." *(A Communication.)*

HENRY KNIGHT, Mechanical Engineer, "Numbering machine, and presses for embossing, stamping in relief in colours, or otherwise."

STANHOPE BAYNES SMITH, Electro-plater and Gilder, "Electro-depositing metals."

THOMAS WALKER, Stove Maker, "Revolving heels for boots and shoes."

WILLIAM BROWN, Machinist, Clerkenwell, London, HENRY SHAPPLE, Child's Hill, near London, Electric Engineer, and WILLIAM WILLIAMS, the younger, Gentleman, "Electric telegraphs and clocks."

THOMAS POTTS, "Suspending curtains, blinds, maps, &c."

DAVID HARCOURT, Machinist, "Vices, machine for dressing hinges and other articles."

BENJAMIN AINGWORTH, Button Maker, "Making ornamental gun skelps."

THOMAS GRIFFITHS, Islington Row, "Tea pots and similar stamped articles."

JOSEPH ROCK COOPER, Gun and Pistol Maker, "Locks for fire-arms."

CHARLES ROWLEY, Button Maker, "Weaver's mails, weaving, dress fastenings, and pins."

JAMES BANISTER, "Boiler tubes."

JOHN COWLEY, Tube Maker (of Walsall), and JOHN HICKMAN, Aston, Clerk, "Metal bedsteads and furniture, telescopic tables, and casting pillars for bedsteads."

RICHARD FORD STURGES, and JONATHAN HARLOW, "Bedsteads."

ENOCH CHAMBERS, Smith, "Wrought iron wheels."

THOMAS WORSDELL, "Envelope making machine."

JAMES GEORGE NEWEY and JAMES NEWMAN, "Buttons, studs, dress-fastenings and ornaments."

SAMUEL FISHER, Engineer, "Railway carriage wheels, axles, springs and hinges for doors."

## 1850.

Robert Harcourt, "Door knobs, window fastenings and fastenings for curtain and other rods."

John Stephen Woolrich (of Wednesbury), Chemist, John James Russell (Handsworth), and Thomas Henry Russell (of Wednesbury), Patent Tube Manufacturers, "Obtaining cadmium from ores."

George Holworthy Palmer, Civil Engineer (of London), and Joshua Horton, Etna Boiler and Gasometer Works, Smethwick, "Gasholders."

William Church, Engineer, "Cutting pasteboard, and numbering and printing railway tickets."

Richard Prosser, Civil Engineer, "Tubes, boilers, and combined steam and hydraulic presses."

Amedee François Remond, "Envelope making machine."

John Turner, Engineer, and Joseph Hardwick, Builder, "Constructing and setting steam boilers."

George Attwood, Copper Roller Maker, "Making tubes of old copper calico-printing rollers."

Abraham Moses Marbe, Chemist, "Hydrocarbons for burning in lamps and for making lacker or varnish, and improved burners for lamps."

Edwin Pettitt, Civil Engineer, "Casting, pressing, grinding, polishing, and working glass."

Jonathan Harlow, "Bedsteads."

Thomas Deakin, Balsall Heath, Gentleman, "Metal tubes, hollow rails, &c."

George Allen Everitt, Kingston Metal Works, Metal and Tube Maker, and George Glydon, Engineer, and Foreman to Messrs. Allen Everitt and Son, "Boiler tubes."

John Manly, junior, "Nails with ornamental heads."

John Percy, M.D., and Henry Wiggin, "Metallic alloys."

Leonard Bower and Thomas Fortune, Harborne, Mechanic, "Screws, bolts, rivets and nails."

Henry Holland, Umbrella Furniture Maker, "Umbrellas and parasols."

Richard Prosser, Civil Engineer, "Feeding boilers and clearing out boiler tubes."

James Henry Williams, "Buttons."

1850 — (*continued*).

WILLIAM LAIRD (of Liverpool), Merchant, and EDWARD ALFRED COWPER, Handsworth, Engineer, "Loading ships with coal and cargo, and modes of building ships."

CHARLES ROWLEY, "Dress pins and other dress fastenings and ornaments."

BENJAMIN HINLEY, Brassfounder, "Castors."

1851.

JOSHUA HORTON, Ætna Works, Smethwick, Boiler and Gas-holder Maker, "Gasholders."

BENJAMIN COOK, "Metallic tubes."

JOHN PERCY, M.D., and HENRY WIGGIN, "Metallic alloys."

GEORGE FREDERICK MUNTZ, junior, "Furnace for making brass."

SAMUEL ALLEN, junior, "Buttons."

JOHN HINKS and JAMES VERO (of Burbage, Leicestershire), "Waterproofing hats, &c."

AMEDEE FRANÇOIS REMOND, Gentleman, "Tubes and gun barrels."

WILLIAM MILLWARD, Plater, "Electro-magnetic and magneto-electric machines."

SAMUEL WALKER, junior, "Metallic tubes."

THOMAS WOODS, Portsea, Upholsterer, and ROBERT WALTER WINFIELD, "Bedsteads, couches, &c."

BENJAMIN WILLIAM GOODE, RICHARD BOLAND, and JAMES NEWMAN, "Chains, chain pins, swivels, brooches, and dress fastenings."

PHILIP WEBLEY, "Boots and shoes."

JOHN FIELDING EMPSON, "Buttons."

JAMES BANISTER, Brassfounder, "Brazing seams of boiler tubes.'

JOHN CHATTERTON, Agent, "Covering insulated telegraph wire with a continuous leaden tube."

JAMES HINKS, "Metallic reels for cotton, &c."

JOHN HOLMES, Machinist, "Cutting-out and stamping metal.'

JOHN ASTON, "Buttons and dress ornaments."

ALEXANDER PARKES, "Separating silver from other metals."

ARTHUR ALBRIGHT, Manufacturing Chemist, "Amorphous phosphorus."

JAMES TIMMINS CHANCE, "Sheet glass."

## 1851 — (*continued*).

TIMOTHY KENRICK, Iron-founder, "Enamelling iron tubes."

JOHN ROWLAND CROOK, Hatter, "Silk hat bodies."

ALEXANDER PARKES, Chemist, "Extraction of copper, and making alloys."

CHARLES GREEN, "Brass tubes."

FREDERICK BENJAMIN GEITHNER, Camden Street, "Castors with china bowls, and making china legs for furniture."

THOMAS BURSTALL, Lee Crescent, Civil Engineer, "Brick-making machine."

JOHN FREARSON, "Cutting out and stamping metals."

JAMES SOUTER and JAMES WORTON, "Papier mâché and buttons made therefrom; buttons and studs of glass and metal combined."

WILLIAM ELLIOTT, "Covered buttons."

## 1852.

RICHARD FORD STURGES, "Ornamenting metallic surfaces by 'nature printing.'"

JOHN HINKS and EUGENE NICOLLE, Civil Engineer, "Nails, bolts, rivets, and screw blanks."

EDWARD SIMONS, Tallow Chandler, "Lamps and candle-sticks."

THOMAS WALKER, "Rotary steam engines."

CHARLES REEVES, junior, "Rolling bayonets, swords, knives, matchets, aud scythes."

ALEXANDER PARKES, "Extraction of silver."

WILLLIAM WESTLEY RICHARDS, Gun Maker, "Revolvers, rifles, gun locks, percussion caps, projectiles, and wads."

ISAAC BROOKES and WILLIAM LUTWYCHE JONES, "Stoves."

JAMES TIMMINS CHANCE, Glass Maker, "Using anthracite coal in glass furnaces."

CHARLES WILLIAM SIEMENS, Engineer, "Fluid meter."

THOMAS ELWOOD HORTON (of Priors Lee Hall, Shropshire), Ironmaster, and ELISHA WYLDE, Engineer, "Boilers, evaporating and heating apparatus."

JAMES STEVENS, Glass Maker, "Lamp glasses."

WILLIAM CHURCH, Civil Engineer, SAMUEL ASPINWALL GODDARD, and EDWARD MIDDLETON, "Fire arms, ordnance, and rifling machine for fire arms."

JOHN HINKS and EUGENE NICOLLE, Civil Engineer, "Plastic compound and machine for moulding the same."

1852 — (*continued*).

GEORGE GOODMAN, junior, "Ornamenting japanned wares and papier mâché."

GEORGE FREDERICK MUNTZ, junior, "Metal tubes."

RICHARD FORD STURGES, "Ornamental wire work fabrics."

SAMUEL LUSTY, "Weaving wire fabrics and making pins."

WILLIAM HETHERINGTON, Gentleman, "Stamping metals."

GEORGE TWIGG, Button Maker, "Buttons and clasps."

GEORGE SHAW, Patent Agent, "Paper bags and envelopes." (*A Communication*).

---

SUMMARY.

1722—1; 1738—1; 1742—1; 1748—1; 1759—2; 1762—1; 1768—3; 1769—2; 1770—2; 1771—1; 1772—2; 1773—1; 1774—1; 1775—1; 1776—1; 1777—1; 1778—3; 1779—4; 1780—5; 1781—4; 1782—1; 1783—2; 1784—1; 1785—3; 1786—8; 1787—6; 1788—1; 1789—1; 1790—7; 1791—1; 1792—3; 1794—3; 1795—3; 1796—5; 1797—1; 1798—6; 1799—2; 1800—5; 1801—3; 1802—3; 1803—1; 1804—4; 1806—3; 1807—2; 1808—8; 1809—6; 1810—9; 1811-11; 1812-11; 1813-10; 1814—6; 1815—2; 1816—5; 1817—6; 1818—5; 1819—8; 1820—7; 1821—6; 1822—3; 1823—7; 1824-12; 1825-12; 1826—9; 1827-11; 1828—5; 1829—9; 1830—9; 1831-10; 1832-12; 1833-12; 1834—8; 1835-16; 1836-17; 1837-16; 1838-22; 1839-24; 1840-33; 1841-34; 1842-29; 1843-22; 1844-24; 1845-29; 1846-30; 1847-22; 1848-23; 1849-30; 1850-22; 1851-30; 1852 (Jan. to Sept.) 21.

Total 776.

---

THE WYATT FAMILY (pp. 7, 21, 86, 87, 121, 160).

Since the foregoing pages were printed, we have received from Mr. Thomas Henry Wyatt, of the War Office (son of the distinguished architect of the same name) some interesting particulars relating to the family. John Wyatt was the eldest of eight brothers, sons of John and Jane Wyatt (*née* Jackson), of Thickbroom, in the parish of Weeford, near Lichfield. John was born in April or May, 1700, his baptism being duly recorded under the date, May 9, in the Weeford parish register. Two of his brothers, Job and William, obtained a patent in the year 1760 for "cutting screws of iron, called wood screws," which is noticed in Mr. Chamberlain's article on the screw trade in "Birmingham and the Midland Hardware District," as being the earliest recorded application of machinery to the manufacture of screws. He had several sons, the eldest being Charles, whose name frequently occurs in the earlier part of this work. His daughter, Cordelia, became the wife of Thomas Dobbs (pp. 28, 88, 156, 157, 169, 177), and their daughter married John Southern, one of Watt's assistants at Soho (p. 28).

Amongst John Wyatt's numerous inventions was one for lessening the friction of bearings by the use of friction rollers. It was not successful, and from a mechanical point of view, it is of no importance whatever; but it seems, nevertheless, to have attracted some attention at the time. It is thus referred to in J. Jacob's "Observations on the Structure and Draught of Wheel Carriages," London, 1773 (p. 53) :—"It is not many years ago that an ingenious artist of Birmingham invented a method to take off the friction from the axles of carriages, from which he expected prodigious advantages in the draught. Nay, he was so confident of success that he caused the words "Friction Annihilated" to be inscribed on the boxes of his wheels. And yet I do not find that his carriage could be drawn faster with the same number of horses than other carriages of the same burthen. It has been said, indeed, that the small rollers on which the axle should have turned did not properly do their office; but if they had, it would have signified little; the diminution of the friction in these cases to more than an eightieth, or at most, an hundredth part of the whole weight of the load, being of no apparent consequence. Before I close this section, however, it may not be amiss to take notice of a late invention, which seems to be an improvement of Mr. Wyatt's scheme above-mentioned; the ingenious improver having, as I have been informed, obviated the difficulties attending flat rollers, by using spherical balls, to prevent the rubbing between the nave and the axle." It was also alluded to by Warltire in a course of lectures delivered by him at Birmingham in the year 1788. Warltire, it may be remembered, was a friend of Dr. Priestley, with whom he had some correspondence on the question of the composition of water. These lectures have never been published, but a MS. note of a portion of the course by an unknown hand, is preserved in the Library of Sir John

Soane's Museum, Lincoln's Inn Fields, London. At page 21 of the volume, the following passage occurs :—"Mr. John Wyatt, of Birmingham, invented a contrivance for that purpose [*i.e.*, the prevention of friction] in blocks of pulleys. In the middle of the pulley, where it ran on the pin, he fixed a box with six small rollers, so that instead of the friction which took place in no small degree between the pulley and the pin, there remained that only of the pivots of the rollers." The reporter then proceeds to describe what appears to be an application of the same contrivance to wheel carriages and the bearings of upright shafts.

The ramifications of the Wyatt family are very numerous, and they offer a singular instance of hereditary talent for art and architecture. It is sufficient to mention that James, President of the Royal Academy, Surveyor General to the Board of Works, Architect to the King, &c. ; Jeffry (afterwards Sir Jeffry Wyatville), who executed important works at Windsor Castle ; Benjamin Dean, Architect of Drury Lane Theatre ; Samuel, Architect of the Trinity House ; Matthew Cotes, the well-known Sculptor ; Sir Matthew Digby, and his brother, Thomas Henry, who died in 1879, were all descended from John Wyatt's brothers.

It is satisfactory to know that John Wyatt's papers, together with the historic hank of yarn spun by the machine under his superintendence, are in the safe keeping of a member of the family, who takes an enthusiastic interest in all that concerns the celebrated inventor.

### Dr. Church.

Dr. Church, who was originally educated for the medical profession, was born in the year 1779 in the State of Vermont, where he died on the 6th of October, 1863, at the house of his son-in-law, having finally quitted Birmingham in 1859. The news of his death does not seem to have reached this country until the 23rd of April following, at which date a memoir of the deceased inventor appeared in the *Daily Post*, and was reprinted in the *Mechanics Magazine* for May 6th. A portrait of Dr. Church, painted by J. Moore, was exhibited at the Royal Academy in 1837, and is noticed in the *Mechanics Magazine*, vol. xxvii., page 119. Enquiries have been recently made as to its whereabouts, but unfortunately without success.

# INDEX.

Simpson, John, 165, 172, 173.
—— Thomas, 53.